T0176899

Supercharg3D

Supercharg3D

How 3D Printing Will Drive Your Supply Chain

LEN PANNETT

WILEY

Library of Congress Cataloging-in-Publication Data:

Names: Pannett, Len, author.
Title: Supercharg3D : how 3D printing will drive your supply chain / Len Pannett.
Other titles: Supercharged | Supercharge 3D
Description: 2nd edition. | Hoboken, New Jersey : John Wiley & Sons, Inc., [2019] | Includes bibliographical references and index. |
Identifiers: LCCN 2018043690 (print) | LCCN 2018044784 (ebook) | ISBN 9781119532361 (Adobe PDF) | ISBN 9781119532385 (ePub) | ISBN 9781119532354 (hardcover : acid-free paper) | ISBN 9781119532361 (ePDF)
Subjects: LCSH: Three-dimensional printing. | Business logistics.
Classification: LCC TS171.95 (ebook) | LCC TS171.95 .P36 2019 (print) | DDC 621.9/88—dc23
LC record available at https://lccn.loc.gov/2018043690

Printed in the United States of America

V10008301_021819

This book is dedicated to my family, for their love, understanding, and patience during the production of this book.
It is especially dedicated to the memory of my father, Bill Pannett, who inspired in so many a sense of wonder in science and technology.

Contents

Preface

THROUGHOUT MY CAREERS, I've worked with clients and organizations from several engineering and technology sectors, and the pressures that they face have all been similar: to meet the particular requirements of their customers as quickly as possible, at the best cost and with the right quality. The Chief Operating Officers (COOs), Chief Supply Chain Officers (CSCOs), and Operations Directors that I've spoken with have all said that those pressures constantly drive them to improve the flexibility, responsiveness, and speed of their supply chains. The successive tweaks they have made in processes and training, however, using tried and tested techniques such as Just in Time (JIT), Kaizen, Lean, and Six Sigma, have delivered diminishing returns. Making a significant change now requires innovation, something truly different. 3D printing can make that difference.

Today's mass media and industry press are full of anecdotes about the wonders of 3D printing. Despite that constant barrage of stories, many supply chain and operations decision makers don't fully understand what 3D printing is, what the technology can and can't do, how it is affecting their supply chain, and how it will do so in the future. Most importantly, they don't understand what they can do about it today: how to consider whether to adopt it, and then how to do so in their business models. Being increasingly tech savvy, they recognize that one can't simply buy a 3D printer and plug it in to produce things that they have made using other methods—there are broader considerations to account for.

This book will help you to make the right decisions and successfully evaluate whether 3D printing is right for your company. It will highlight what 3D printing can do for you and your customers, and how to adopt this disruptive technology. By reading this book, you will discover how to drive value in your supply chain by supercharging it with 3D printing.

Preparing a text on the impact of a fast-changing technology on business operations is a daunting task. It can easily become too technical and hence quickly outdated. Alternatively, it can be too basic, leaving doubt and confusion in the mind of the reader. It can also be written at the wrong time, either too early to be relevant or too late to be useful. I have thus prepared this book by taking a balanced view, presenting sufficient technical information on 3D printing without losing focus. Moreover, the feedback I have received from clients, peers, and industry observers indicates that it is needed now—that 3D printing has to be considered now in the context of how supply chains operate. While technological developments may move along quickly, the essence of how 3D printing will drive supply chains as described here will hold for a while yet.

Acknowledgments

THIS BOOK WOULD not have been possible without the help, advice, and support of many colleagues and friends.

Particular thanks to Arvid Eirich and Stephanie Brickwede at Deutsche Bahn; to Phil Reeves at Stratasys, Marc Saunders at Renishaw, Nick Lewis at 3DSystems, and Dion Vaughan at Metalysis; and to Alan Amling at UPS, for all their insights into the realities of 3D printing. Thanks are also due to Michael Petch for his early guidance.

I thank Cynthia White for her guidance on how to steer the narrative; Beth McLoughlin for her advice on writing style; and Adam Brocklehurst, Andy Hindle, and Frank Kaye for their patience in reviewing the content. I'd also like to thank Margaret Cummins, who believed in the book from the off, and the team at Wiley for their support. Special thanks to Gavin Chan for helping me get to the right desk in the first place.

Thank you also to Robert Dudley for trusting in a new author who had some set ideas.

I especially thank my wife, Lorna, for her patience and guidance in endless late-night discussions.

Introduction

AT 05:52 UNIVERSAL TIME on September 21, 2014, the company SpaceX launched its sixth resupply mission, SpX-4, toward the International Space Station (ISS), under a partnership with NASA. Arriving two days later, it was received by the crew of Expedition 41/42, which had already spent four months there in isolation. On that mission was a first-of-its-kind piece of equipment: a new technology that would allow the ISS crew to carry out in-orbit repairs themselves without needing to wait for the resupply of components from the ground. Built by the company Made in Space, only founded four years earlier, the so-called 3D Printing in Zero-G experiment placed a 3D printer on the ISS to test how the technology would behave in microgravity. Installation of the device was completed on November 17, and a calibration test conducted three days later, which indicated the need for minor modifications that were transmitted from the ground control team and resolved remotely.[1]

On November 24, the device was sent a digital file containing the design specifications of a new faceplate for the printer, an item that was damaged during the 400 km journey up. The next day, the finished part was removed from the printer, inspected by NASA astronaut Barry "Butch" Wilmore, and installed.[2] Here was a new technology that allowed parts to be made locally, remote from the normal physical supply chain, in perhaps the harshest of environments. It allowed a team of designers and engineers to send a digital file to a printer installed hundreds of miles away, without the need for the people at the destination to do any manufacturing themselves—a highly symbolic moment in the development of this transformative technology.

Many industry sectors today—particularly, but not exclusively, manufacturing, engineering and technology—already have some level of 3D printing in their operational processes. Historically, the architectural and automotive sectors have been the greatest users of this technology, as an aid for prototyping and building models. The medical and dentistry sectors have increasingly employed it to make products to be used with patients directly or to help them prepare for procedures. Ninety-nine percent of the world's in-ear hearing aid shells are custom made and 3D printed,[3] and today over 10 million such devices are in use worldwide, produced by companies like Swiss manufacturer Sonova. Indeed, in the United States, the outer shell of every in-ear hearing aid used today is 3D printed, and the entire American sector converted to that technology in under 500 days.[4] The jewelry and hobby craft industries are increasingly adopting 3D printing, as is manufacturing, from high-tech sectors such as aerospace to older industries like mining, transport, and rail. In 2016, manufacturers sold over US$2.5 billion in 3D printers, rising at a rate of 15–20% annually. The value of products and parts made using the technology is already in the range of several dozen billions of dollars, with estimates of the potential size of the market ranging from Boston Consulting Group's figure of US$350 billion in 2035,[5] to the more aggressive estimate from their rivals at consulting firm McKinsey of US$550 billion by 2025.[6]

Over the past 10 years, 3D printing has featured more and more in the public eye. Driven by what has been at times a media frenzy, a steady stream of good news stories has been published, each describing some new application, from helping a blind mother-to-be feel what her fetus looks like during her gestation, or providing a child born without a nose with a 3D-printed plastic prosthetic, to the manufacture of entire houses using mammoth-sized printers. Many of these reports contain a large measure of hype, with promises that soon we will have *Star Trek*–like replicators in our homes to make what we want, when we want it, how we want it. Indeed, the hype became so big that a new entrant to the market, MakerBot, which sold simple and cheap 3D printers for the home market, quickly overtook long-established firms such as 3D Systems and Stratasys in terms of the number of machines sold.

The reality, though, is more conservative. 3D printing is a long way from the plug-and-play scenarios that much of the excitement described, and those who have tried 3D printing in their homes have been largely disappointed and frustrated with what it can do. However, when its industrial uses are considered, the potential of 3D printing is astonishing. Much as personal computers, cell phones, and the Internet have done, 3D printing is quite simply a disruptive technology, a fact highlighted by President Barack Obama in his 2013 State of the Union address:

> "3D printing […] has the potential to revolutionize the way we make almost everything."[7]

This revolution is already changing how things are made, much as other disruptive technologies have done, opening up new ways of working and enabling new supply chain models. While President Obama may have been a little optimistic about the extent to which 3D printing will disrupt manufacturing—it is unlikely to replace all manufacturing—it is fast expanding its presence in production. In 2016, Dick Elsy, Chief Executive of the UK's High Value Manufacturing Catapult, said that 3D printing has:

> "Enormous potential which, when fully realized, will transform product development, supply chains and manufacturing as we know them."[8]

Indeed, it will surprise many that they are already using products that have involved 3D printing at some point in their development and manufacturing lifecycle. For instance, the aircraft maker Airbus's new wide-body A350 XWB aircraft uses 2,700 parts created in plastic using 3D printing, and the company is working with the European Aviation Safety Agency (EASA) to qualify titanium components produced on 3D metals printers.[9] By mid-2018, the company had orders for over 882 A350XWBs from over 46 customers worldwide.[10]

3D printing already brings significant advantages to the different elements of the supply chain, and more will emerge. From accelerating product development from months and years to days and hours, to reducing delivery lead times to a matter of hours, 3D printing is already changing the dynamics, size, and shape of supply chains in many sectors. However, several technical and commercial hurdles must be overcome for it to make the significant contribution its advocates promise. The good news is that these barriers are recognized by those in the 3D printing ecosystem and progress is being made to overcome them. With improvements in the quality, accuracy, and precision of 3D printers, and the emergence of new commercial models, the industry is advancing like never before. Indeed, those improvements have been driving the growth in the use of 3D printing across industry sectors.

In 2012, around the time that MakerBot was overtaking the stalwart companies in the sector, industry observers, who annually map the hype surrounding emerging technologies, identified 3D printing as being at the height of expectations. The share prices in the leading 3D printing companies increased significantly, and later, in 2013, in a reported US $604 billion deal (with $403 million of that up-front in MakerBot stock),[11] MakerBot was acquired by Stratasys—by then the largest and one of the oldest 3D printer makers. However, there was a growing realization that the aspirations for 3D printing far outstripped the reality and public disillusionment increased, with an impact on sales of consumer 3D printers, that is, those aimed at 3D printing in the home, and share prices of the key players. At the heart of that was the relative difficulty and inconvenience to design an item, the restricted range of materials, the time it took to make something, and the cost of materials.

While consumer 3D printing stalled, the situation was far more nuanced for industrial uses in 2013. It was long a mature technology for prototyping, ever since the Ford Motor Company first acquired a 3D printer in 1986 and ancillary technologies such as 3D scanning and 3D design software were fast becoming common tools in enterprises. Much as consumer 3D printing had been hyped, its use in manufacturing was likewise entering a period where expectations surpassed reality. Driven by increased competition and market pressures, though, this role quickly grew and by 2015 it was well on its way to entering operations models as businesses recognized its potential. As patents expired and as technologies evolved, so the costs of industrial 3D printing decreased while its performance increased. With that came a step change in the number of companies making 3D printers; from a small handful 15 years ago, to today, with 3D printing as one of the fastest-growing sectors, there are more than 100 new printer makers, with an expanded number of software developers and firms producing raw printer materials. This growth has made the technology viable in industrial settings where before it was priced out or not fit for purpose, too restricted in what it could do.

In parallel with the technological changes, supply chains themselves are now providing fertile ground for the adoption of 3D printing. Consumers increasingly seek personalized end products, driving the need for a "customization of one," be it for mobile phone covers, sports shoes, or cars. Industrial customers want solutions to their material needs that are quicker to deliver and, likewise, better fitting their specifications. Moreover, cycle times for new designs are shrinking, as customers across industries demand new features and new applications be made available as quickly as possible. Facebook, for instance, has introduced regular updates to its websites to appear always fresh, and Apple has built a cult-like following as it releases new models of its iPhone and iPad product families every year or two. Concurrently, customers are increasingly intolerant of errors in delivery, constantly advocating for the right thing to arrive at the right time, at the right price. Put together, those factors are driving businesses to respond to changes in demand and designs, and deliver those "new and improved" products to their customers wherever they may be, right the first time. It is here that the advantages and capabilities of 3D printing come to the forefront, both to meet consumer wants and industry customer needs. With supply chains now true differentiators of business success, companies that adopt 3D printing in their supply chains will be at an advantage.

This is changing the concept of value chains—how things get made, from an idea to a finished product in the hands of the customer—and therefore the models that enable them. Traditionally, the things we make and use are produced from raw materials and components (themselves made of raw materials and components), which are brought together and assembled into the finished product. Throughout that process, those raw materials, components, and unfinished and finished items are typically stored until they are needed in the next step of the value chain (Figure I.1).

3D printing is changing this paradigm, allowing for things to be made with fewer raw materials, with fewer parts and closer—or even inside—the next step of the chain, eliminating the need for storage and distribution and shortening supply chains (Figure I.2).

The trend is already accelerating, and if a company is not now considering what to do about 3D printing, they have already fallen behind their customers and competition. Those firms now considering how 3D printing will affect their supply chains, and the benefits (and challenges) that it brings, and making the necessary adjustments to their operating and supply chain models, will lead tomorrow's commerce. As Dr. Phil Reeves, Vice President at Stratasys Consulting says, "Engineers understand what 3D printing is. Now there is a need to get the commercial areas, such as supply chain, to understand how to exploit it."[12]

This book aims to help with that.

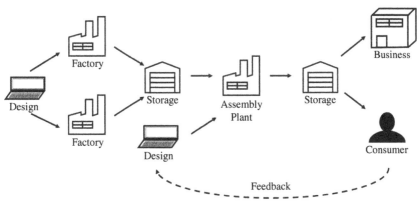

FIGURE I.1   The traditional value chain.

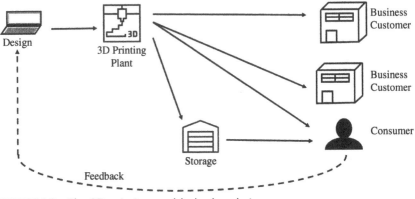

FIGURE I.2 The 3D-printing-enabled value chain.

THIS BOOK

A disruptive technology is one that completely changes the way in which companies work, how people interact with those companies and its products, and how they interact with themselves. The last 100 years have seen Henry Ford's model of mass production, the invention and use of computers in business and, more recently, innovations such as the Internet, the World Wide Web, and smartphones, all of which have changed how we think, work, and play. 3D printing is certainly such a disruptive technology. Most of the commentary about the technology itself—the different methods of 3D printing, the materials it uses, and the software packages called for—eventually says something about how the technology is changing the way things are made, even going as far as to predict timelines. Very few of those commentators take a hard look at how 3D printing will change how businesses work, both internally and with one another. It is time to address that missing piece in the narrative.

Increasingly senior managers are asking four questions:

- What is 3D printing about?
- Why should I care about it in my company?
- Should I use it in my firm?
- How do we go about adopting it?

This book will guide the reader through these, equipping them with the information they need to answer them for their particular company. It will do so by taking a pragmatic position, balancing the opportunities that 3D printing presents with the reality of the limitations that it continues to have. Some readers may already be aware of the technology and what it can do, while others will be approaching this without that knowledge. Still others already may be contemplating how to adopt 3D printing, certain that there would be benefits to them and their customers but unsure how best to proceed. This book has been structured so that each of them can find what they need, whether they read it from cover to cover or jump straight to the relevant chapter without the preamble. The aim is for all to take away the information they need from this book, to return to it as an aide-mémoire, scribbling notes in its margins and marking pages as needed.

This book will show how 3D printing is already part of many supply chains and that this will increasingly be the case. It starts in Chapter 1 with a look at the basics of 3D printing technology, providing the necessary background information for understanding what it is, how it works, and what it can and cannot do. However, it stops short of offering a detailed survey of current technology or an in-depth description of the capabilities, advantages, and constraints of each one; attempting to do so would be fruitless, as the pace of change in 3D printing technology quickly renders such narratives out of date. Instead, it will provide sufficient detail for readers to appreciate these parameters. If more is required, the reader is encouraged to seek other channels for the latest developments, such as the excellent blog http://3dprintingindustry.com, specialist magazines such as *Additive Manufacturing* and *TCT*, and other manufacturing technology publications. Chapter 2 describes the features and capabilities of 3D printing, what it can and can't do, and its benefits and constraints, brought to life with contemporary examples from supply chains across different sectors. These include issues of speed, materials, design and design tools, volumes, finish, costs, locations, labor, and sustainability. Chapter 3 then looks at where 3D printing technology is heading, not just in isolation but also more widely as part of the broad range of digital advances that are increasingly transforming supply chains. Chapter 4 identifies the channels that supply chains can use to access 3D printing, from acquisition and leasing of equipment to outsourcing to specialist third parties. Chapter 5 focuses on a specific case to bring this to life, using the experience of the German transport firm Deutsche Bahn to tell the story of how 3D printing can drive supply chains. Chapter 6 then breaks the supply chain into its individual elements, using the SCOR® model as a basis, analyzing how 3D printing affects each of the Plan, Source, Make, Deliver, Return, and Enable. Chapter 7 examines the current and emerging supply chain models that 3D printing enables, from producing things in-house and manufacturing at customer sites, to enabling customers to make their own things with suppliers' designs. Chapter 8 takes a closer look at the wider implications of using 3D printing that supply chains need to consider, including legal, quality management, standards, regulation and accreditation, health and safety, data management and security, commercial models, fiscal and financial impacts, and skills. Chapter 9 presents the case that, in the face of the changes that 3D printing is already bringing, businesses should actively consider how it will affect them and what they can do about it. It will look at how manufacturing, engineering, and technology companies can analyze their supply chains, and answers the questions "How do we analyze whether to adopt 3D printing in our supply chain?" and "How do we go about adopting it?"

 ## CONSUMER VERSUS INDUSTRIAL 3D PRINTING

Much of the narrative on 3D printing concerns how it will change people's day-to-day lives of people, how we will all have 3D printers in our houses, on our desks, to replace everything from drawers of odd screws and washers to the latest, customized gadget, just like those aforementioned replicators in the *Star Trek* tradition. All of those uses are consumer 3D printing. Even if we have a revolution in the skills, technologies, and systems involved to make that happen, thus allowing everyday folk to design and print on demand, that future is decades away.

Instead, this book will concern itself with the short- and medium-term situation in an industrial environment, rather than examining the consumer 3D printer market, leaving the debate over that area to other authors. Throughout the book, the terms "customer" and "end user" will be used to refer to those who use the items and products that 3D printing makes in an industrial context (i.e. those that are provided by industrial businesses, from small or medium enterprises to large, global conglomerates).

 ## 3D PRINTING VERSUS ADDITIVE MANUFACTURE

When the techniques that became what we know as "3D printing" began to emerge in the 1980s, they were primarily aimed at making test objects, establishing the concept of "rapid prototyping." It was one of a series of technologies, tools, and techniques that aimed to make prototypes far more quickly than previously achievable, and it was used by architects to make models of buildings and by automotive designers to produce mockups of new cars and auto parts. As the technologies matured and started to move toward industrial uses in manufacturing, they were labeled "additive manufacturing," a term that described the process literally, one involving making things by adding raw materials rather than removing them from an initial volume, which itself came to be known as "reductive" or "subtractive manufacturing." When the international standards organization ASTM International composed the first Standard Terminology for Additive Manufacturing Technologies in 2012, it defined additive manufacture as "a process of joining materials to make objects from 3D model data, usually layer upon layer, as opposed to subtractive manufacturing methodologies."[13] (This was slightly amended in the 2015 nomenclature to include "formative technologies," such as forging, rolling, and sheet metal working.) This was differentiated from 3D printing, defined therein as "the fabrication of objects through the deposition of a material using a print head, nozzle, or another printer technology."

The expiration of some key patents and the rapid increase in the number of companies making the tools that did additive manufacturing saw these firms enter the public arena through a rapidly rising number of media articles, feeding the hype in the early twenty-first century while causing share prices of the biggest makers of those machines to rocket. However, many journalists thought the term "additive manufacturing" too obscure and preferred "3D printing" instead, as it was easier for an uninformed person to understand. Those directly involved in the 3D printing industry initially reserved the term "3D printing" for the consumer end of the market, encompassing companies like MakerBot, which made the US$1,000 machines sold in retail outlets like Staples and Office Mart. Over the last few years, as the existing technologies involved have become more widely known and as new ones have emerged, and as the machines involved have become more competitive in prices and end-to-end costs, the term "3D printing" has become more generalized, and it is usually used synonymously with "additive manufacture," something that the ASTM standard first noted in 2012 and again in its revamped 2015 ISO/ASTM 52900:2015 version: "[3D printing is] often used in a non-technical context synonymously with additive manufacturing."[14] The issue of whether the set of technologies is called "additive manufacturing" and/or "3D printing" continues to be debated, with many good arguments on both sides; those will not be narrated in this book. At their hearts, both names refer to making things by rendering raw materials layer by layer. For the sake of simplicity and clarity, this book will use the term "3D printing" to cover both throughout.

THE CONCEPT OF THE SUPPLY CHAIN

As it is a core element of this book, it is important to clearly understand the concept of the supply chain. All too often, supply chains are interpreted as referring solely to logistics, warehousing, or purchasing. Much effort has been expended by authors, consultants and academics to define what a supply chain is. Taking the cue from them, this book will use the SCOR definition: a supply chain consists of planning, sourcing, making, delivery, and returns. This is the industry-standard model that most business leaders readily understand, and it covers all relevant elements both within and between companies. Where appropriate, this book describes the parts of the SCOR model in a little detail, but the reader is referred to the many external sources on the model for further edification, such as the American Production and Inventory Control Society (APICS) websites and textbooks.

Notes

1. Q. A. Bean, K. G. Cooper, J. E. Edmunson, M. M. Johnston, and M. J. Werkheiser, "International Space Station (ISS) 3D Printer Performance and Material Characterization Methodology," 2015, https://ntrs.nasa.gov/archive/nasa/casi.ntrs.nasa.gov/20150016234.pdf.
2. NASA, "Open for Business: 3-D Printer Creates First Object in Space on International Space Station," November 25, 2014, https://www.nasa.gov/content/open-for-business-3-d-printer-creates-first-object-in-space-on-international-space-station.
3. P. Reeves, *Additive Manufacturing & 3D Printing Medical & Healthcare: A New Industrial Perspective* (Derbyshire, UK: Econolyst, 2014).
4. Richard D'aveni, *The 3-D Printing Revolution* (Boston: Harvard Business Review, May 2015): 40–48.
5. D. Küpper, W. Heising, G. Corman, M. Wolfgang, C. Knizek, and V. Lukic, "Get Ready for Industrialized Additive Manufacturing," Boston Consulting Group, April 5, 2017, https://www.bcg.com/en-gb/publications/2017/lean-manufacturing-industry-4.0-get-ready-for-industrialized-additive-manufacturing.aspx.
6. D. Cohen, M. Sargeant, and K. Somers, "3D Printing Takes Shape," *McKinsey Quarterly*, January 2014, https://www.mckinsey.com/business-functions/operations/our-insights/3-d-printing-takes-shape.

7. Barack Obama, "Remarks by the President in the State of the Union Address (February 12, 2013)," https://obamawhitehouse.archives.gov/the-press-office/2013/02/12/remarks-president-state-union-address.

8. D. Elsy, "Viewpoint: Catapult Chief Dick Elsy on the Government's Additive Strategy," The Engineer, October 13, 2016, https://www.theengineer.co.uk/viewpoint-catapult-chief-dick-elsy-on-the-governments-additive-strategy.

9. Aerospace Manufacturing, "Adding Value with 3D Printing," https://www.aero-mag.com/airbus-a350-xwb-a320-neo-3d-printing-aerospace-sector-stratasys-ultem-9085-fused-deposition-modelling-fdm/.

10. David Casey, "Farnborough Airshow 2018–Latest News and Order Updates," Routes Online, July 20, 2018, https://www.routesonline.com/news/29/breaking-news/279602/farnborough-airshow-2018-latest-news-and-order-updates/.

11. Kelly Clay, "3D Printing Company MakerBot Acquired in $604 Million Deal," June 19, 2013, https://www.forbes.com/sites/kellyclay/2013/06/19/3d-printing-company-makerbot-acquired-in-604-million-deal/.

12. Interview with the author, July 2017.

13. "ASTM F2792-12a. Standard Terminology for Additive Manufacturing Technologies (Withdrawn 2015), ASTM International," accessed June 30, 2017, www.astm.org.

14. "ISO/ASTM 52900:2015. Additive Manufacturing—General Principles—Terminology," accessed June 30, 2017, https://www.iso.org/obp/ui/#iso:std:iso-astm:52900:ed-1:v1:en.

CHAPTER ONE

What Is 3D Printing?

ISTORY IS REPLETE with examples of technologies that were invented long before they were widely used. Ethanol fuels were used to power Model T Fords in 1908, many decades before they transformed the automotive sector in Brazil in the 1970s and the rest of the world thereafter. The Internet was created in the late 1960s, 30 years before it entered mainstream business and changed so much of what we do. 3D printing is another one of those technologies. For millennia, making something involved taking materials away from an initial quantity, whether it was whittling a spear from a stick, carving a marble block into the form of Hercules, or producing an intricate table leg on a lathe. Those techniques are all "reductive" because they removed material from an initial volume, and today include drilling, milling, cutting, and molding materials into the final required form latterly termed "formative" manufacturing. Then in 1984 that all changed.

It will surprise many who might be unfamiliar with 3D printing to find that it is over 30 years old. In 1980, when companies had only just started using desktop computers to help with engineering and business processes, Dr. Hideo Kodama of the Nagoya Municipal Industrial Research Institute in Japan (Figure 1.1) developed an idea: to expose a vat of photosensitive resin to an ultraviolet (UV) beam that hardens a layer of the illuminated liquid,

FIGURE 1.1 Dr. Hideo Kodama, Nagoya Municipal Industrial Research Institute in Japan (Image courtesy of IEICE).

and then to build up the layers to form a solid object. He applied for a patent for his idea in May 1980, but the full specification wasn't completed within the filing deadline because of a lack of funding. He submitted what became a fundamental paper, titled "A Scheme for Three-Dimensional Display by Automatic Fabrication of a Three-Dimensional Model" to the Institute of Electronics, Information and Communication Engineers (IEICE) in Japan, which was published in its journal *Transactions on Electronics* in April 1981.[1]

Later that decade, an engineer working for UVP Inc. in San Gabriel, California, a small firm that placed layers of veneer on tabletops and rubber tiles, thought there might be some interesting possibilities by doing things differently. Charles "Chuck" Hull—the so-called Thomas Edison of 3D printing—was their Vice President of Engineering and oversaw the process, which used a UV laser beam to turn acrylic photopolymers from liquid to a plastic-like solid. Thinking about the process, it occurred to him that the same machines that did the layering could form the resultant plastic into any shape that he liked and, by building these layers one on top of the other, he could make a three-dimensional object. On March 9, 1983, having spent some time developing the code to instruct the machine in what to do and then setting it to work, he phoned his wife and called her to his workshop. "You've got to see this!" he said excitedly.

With a response that "it had better be good," she took the short trip to the office and there Chuck presented her with his creation: a small eye-wash cup made of black plastic, the shape chosen mostly because it was easy to make. The machinery itself wasn't particularly elegant, Chuck himself recalls:[2]

> "[The 3D printer] was so kludged together that it looked postapocalyptic, like some of the equipment they used in that movie *Waterworld*.

So was born *stereolithography* (SLA) in 1983—a term made up from three Greek words, "stereo-" meaning solid, "litho-" meaning rock or stone, and "-graphy" meaning writing; in combination, "writing solids in stone"—the first *additive* manufacturing technique, whereby three-dimensional objects are made by building them layer upon layer. The equipment was too heavy and fragile to lug around, so he took a video of the process and showed it to several executives across the United States. In August 1984, Chuck submitted his patent, titled "Apparatus for Production of Three-Dimensional Objects by Stereolithography," which was granted on March 11, 1986* and he set up the first dedicated company, 3D Systems, in Valencia, California (and later moved to Rock Hill, South Carolina), to commercialize it, subsequently raising US$ 6 million from investors.

3D Systems first commercial product was released in 1988 and car manufacturers immediately noticed; famously, the Ford Motor Company bought the third 3D printer ever made. At the time, making a prototype using conventional means, including the necessary tooling up and production, took some six to eight weeks, so having a machine that could make one in a matter of hours—albeit crudely—was a revolution. 3D Systems went on to be the leading company in the sector for the next 15 years. Several other techniques have since emerged, all based on the concept of building objects layer by layer, and the pace of development has accelerated markedly in the last few years.

Understanding what is meant by 3D printing is the first step to exploring what it can bring to a company. However, with today's plethora of different technologies and myriad variations of those, the 3D printing industry makes

---

* As is the case with many significant inventions, the realization of 3D printing started with two parallel developers, but only one is today called the original inventor. The first patent for stereolithography was actually submitted by Alain Le Méhauté, a scientist and engineer-in-chief at General Electric's Research Centre, together with Olivier de Witte and Jean Claude André, on July 16, 1984, three weeks before Chuck Hull. However, their application was abandoned "for lack of business perspective" by the sponsoring organizations involved. See A. Moussion, "Interview d'Alain Le Méhauté, l'un des pères de l'impression 3D," September 17, 2014, http://www.primante3d.com/inventeur/.

this a difficult task. Technologies are grouped under different headings, often using conflicting grouping schemes, further confusing the matter. Moreover, the research and development departments of universities, 3D printer makers, and raw materials producers are constantly adding to those new approaches. The result is that navigating the sea of what 3D printing is can be a stormy journey.

This chapter aims to cut through the terms and provide a measure of clarity. It provides an overview of what 3D printing is, looking at the main techniques used at present. However, it is not to be used as an authoritative reference on 3D printing technology, as its fast-changing nature would render this book out of date before it was published. Rather, it will provide the reader with sufficient detail to understand how things are made using the different types of 3D printing, knowledge that will be needed to assess what it can be used for and how to use it. It will also provide an outline of the end-to-end 3D printing process common across all techniques.

 ## 3D PRINTING TECHNIQUES

The early patents in 3D printing technologies started to expire at the turn of the twenty-first century, fueling an explosion in start-ups and global companies' R&D divisions seeking new ways to achieve similar ends: the making of three-dimensional objects using additive techniques. The pace of development has been breakneck. According to John Hornick, a partner at law firm Finnegan IP and a 3D printing specialist, 109 patents related to 3D printing were issued by the US Patent Office in 1999, covering materials, equipment, and software; by 2008, that number had increased to 186, and by 2016 to 646, with a further 1,842 patent applications submitted.[3] By definition, patents require a "unique" approach, but 3D printing today predominately rests on one of seven foundational methodologies (Table 1.1):

- vat photopolymerization
- material jetting
- binder jetting
- material extrusion
- powder bed fusion
- direct energy deposition
- sheet lamination

TABLE 1.1 Summary of 3D printing techniques.

Technique	Variation
Vat Photopolymerization	Stereolithography (SLA)
	Digital Light Processing (DLP)
	Continuous Digital Light Manufacturing (cDLM)
	Continuous Liquid Interface Production (CLIP)
Material Jetting	Material Jetting (MJ)
	Nanoparticle Jetting (NPJ)
	Drop on Demand (DOD)
Binder Jetting	Binder Jetting (BJ)
Material Extrusion	Fusion Deposition Modeling (FDM)
	Fused Filament Fabrication (FFF)
Powder Bed Fusion	Multi-Jet Fusion (MJF)
	Selective Laser Sintering (SLS)
	Selective Laser Melting (SLM)
	Direct Metal Laser Sintering (DMLS)
	Electron Beam Melting (EBM)
Direct Energy Deposition	Laser Engineering Net Shape (LENS)
	Electron Beam Additive Manufacturing (EBAM)
Sheet Lamination	Laminated Object Manufacturing (LOM)
	Ultrasonic Additive Manufacturing (UAM)

Vat Photopolymerization

Vat photopolymerization uses UV light to selectively harden photopolymers or epoxy resin into a solid layer before then hardening the next layer. Chuck Hull's original invention took a plastic resin and created a 3D object by building it up layer upon layer across a flat plane using a low-powered UV laser to solidify the resin in the required shape. This technique (Figure 1.2), SLA, became the prevalent method for 3D printing for the first few years. A similar approach is Digital Light Polymerization, which was developed by Al Siblani and Sasha Shkolnik in the late 1990s and uses a light projector instead of a laser. The company they formed on the back of Digital Light Processing (DLP), EnvisionTEC, went on to develop DLP into another process, Continuous Digital Light Manufacturing (cDLM). That approach sees the build plate—the plate upon which a 3D-printed object is made—moved continuously with the goal of producing items much faster than DLP.

In 2013, the start-up Carbon3D demonstrated a new photopolymerization-based technique, Continuous Liquid Interface Production (CLIP), which produces objects in a vat of resin by building the layers top-down, using a mixture of light-emitting diodes (LEDs) and oxygen inhibition to harden the

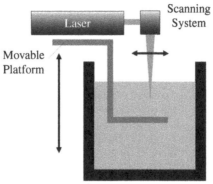

FIGURE 1.2　SLA 3D printing.

photopolymers at the bottom of a vat (Figure 1.3). After a layer is solidified, it is raised, and the next layer is hardened to the bottom of the previous one; when viewed, the overall effect evokes images of the liquid/solid T-1000 Terminators seen in James Cameron's movies. Early demonstrations of the technology, notably by vehicle manufacturer Ford, have found that CLIP is not only fast but also produces stronger items than other photopolymerization techniques.[4]

Clearly, as a resin is needed, the range of materials that can be used is restricted to those which are plastic and plastic-based.

A major advantage of vat photopolymerization is the high degree of accuracy that the technique allows for, producing good finishes. DLP, for example, was quite successful in the jewelry sector because it is precise and produces a good surface finish. Another plus is that, compared to other techniques, vat

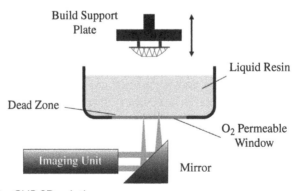

FIGURE 1.3　CLIP 3D printing.

polymerization is relatively swift, producing items in a fraction of the time of other methods; CLIP in particular can produce objects that would take other techniques several hours to complete in under an hour. Moreover, the technique can make items with large build areas (up to a square meter) and weights (up to 200 kg).

Compared to other techniques, however, vat polymerization is expensive, with resins accounting for much of that higher cost. It also has some clear restrictions: as objects are made within the liquid resin, the dimensions, weight, and center of gravity of the fabricated item must be considered if the item is not to topple during the 3D printing process, requiring structural supports to be added to the design of the object being made. These supports will then need to be removed after fabrication using cutting or filing. As with many other 3D printing techniques, vat photopolymerization process requires considerable postprocessing time to cure the materials fully, and the designs employed must consider the removal of resin as the fabricated items are made.

Material Jetting

Imagine taking a printed sheet of paper from a normal 2D inkjet printer, putting it back into the printer to place another layer of print on top of the dry ink, and repeating that hundreds if not thousands of times, building up the layers of ink to form an object. That is material jetting (Figure 1.4), whereby material is placed onto a base by a print head, either in a continuous jet or at discrete locations, the latter being termed Drop on Demand (DOD). Once placed, the material then solidifies, either by itself or by using ultraviolet (UV) light to catalyze the process.

This technique uses similar materials to SLA, primarily thermoset photopolymer resins (i.e. they harden when light is applied to them) with the right

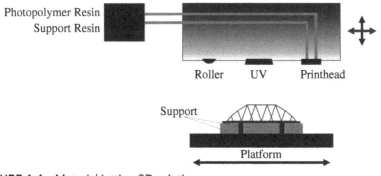

FIGURE 1.4   Material jetting 3D printing.

physical properties for jetting and curing, in terms of viscosity and the ability to form droplets. This includes the more common forms of those materials like polypropylene, making material jetting more suitable for low-cost 3D printing. New variations, such as XJet's NanoParticle Jetting (NPJ), enable metal and ceramics to be used.

Material jetting techniques can be used to create reasonably smooth-surfaced objects with complex designs. They have become popular in the manufacture of injection and casting molds in medicine and dentistry, and in the craft industries such as jewelry. The accuracy of the material jetting can be very high, particularly when DOD is used, which reduces materials wastage. Moreover, the technique allows for several materials—and therefore colors—to be included in a single fabrication run. As the technique makes things bottom-up, geometries and designs need to account for a lack of support structures, although these can be temporarily placed on the print bed and removed later.

Binder Jetting

Binder jetting creates items by using a chemical reaction between a powder-based material and a binding agent, the latter usually in liquid form (Figure 1.5). To make an item, a print head deposits powder as it moves along the printer's x- and y-axes, repeating that journey with a binder that is only placed at the locations that make the intended object. The entire bed is then lowered a little and the process repeats itself.

For a considerable time after it was developed, this technique could only use specialist plastics, although more recent versions have introduced stainless

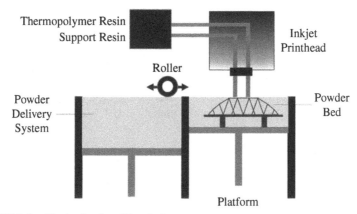

FIGURE 1.5 Binder jetting 3D printing.

steel and some ceramics, including glass, to the family of usable materials. By employing different print heads simultaneously, binder-jetting-based 3D printers can allow for parts to be made with several different colors and even materials at the same time, something that increases its range of uses. The process is reasonably fast and, by judiciously employing different powder–binder pairings, a range of mechanical properties can be produced in an item. However, this process is not generally suitable for items destined for situations where stresses to structural strengths prevail. Moreover, the tolerances that the technique allows for mean that postprocessing can be significant, adding to the overall process time.

Material Extrusion

Rather than using resins, extrusion-based techniques use melted thermoplastics (i.e. plastics that are soft when heated and hard when subsequently cooled) as filaments that are squeezed through heated nozzles in stripes or "pipes" of material. Objects are made on a plate by moving the nozzle on the x- and y-axes and lowering the plate for the z-axis, building objects from the bottom-up. The materials can be bonded by controlling temperature or using a bonding agent. As a result, if there are overhanging parts in the design, supports must be included that are removed after the item is made, as with vat photopolymerization and material and binder jetting.

The first extrusion technique to be commercialized was Fusion Deposition Modeling (FDM), developed by S. Scott Crump in the 1980s. He got the idea for it while he was making a toy frog for his two-year-old daughter using a glue gun with polyethylene and candle wax as he sat in his kitchen. Like Chuck Hull, he also saw the potential to make objects by building them layer upon layer, having been inspired by the movements of plotter pens, which draw architectural and mechanical blueprints on large sheets of paper. At the urging of his wife, Lisa, they started Stratasys in Eden Prairie, Minnesota, and patented his idea for FDM in 1989. When that patent expired in 1992, a wide community of developers began to work with it, later relabeling it Fused Filament Fabrication (FFF) to avoid conflicts with the progenitor technology. FDM (Figure 1.6) is relatively cost-effective and is typically used for making prototypes quickly and for low-cost product development. As advancements in materials science have provided improved plastic base materials—the only set of materials this technique can use—materials extrusion has moved into the manufacture of items destined for end uses.

As well as being reasonably inexpensive and widespread, the items made have properties that make it useful for making prototypes and for

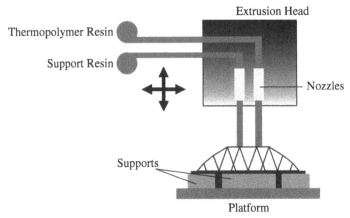

FIGURE 1.6   FDM 3D printing.

low-cost production. However, the physical limitations of materials extrusion equipment limit the quality of what is made. For instance, nozzles must have radii that allow them to release materials that will bind to laid-down layers. This also impacts the accuracy of what is made, which not only limits the purposes for which these techniques can be employed, but also means that postproduction and finishing is needed to yield the higher qualities that end-use objects typically must exhibit.

Powder Bed Fusion

The Powder Bed Fusion (PBF) family of techniques is commonly used for industrial parts, particularly those destined to make metal objects. It was initially developed in 1984 by Carl Deckard, a mechanical engineering graduate at the University of Austin in Texas in his senior year. He spent two years refining his idea, something that led him to pursue a master's degree under the tutelage of Dr. Joe Beaman, finally developing a contraption that was baptized "Betsy." When it operated, Deckard used a large salt cellar to dust a surface with a thin layer of powdered plastic, which was then heated by a 100W laser guided by a supercharged Commodore 64 personal computer.[5] The laser traced out a desired shape in the plastic powder, melting this where it was lit. Another layer of powder was then sprinkled over the entire bed and the sequence repeated, building up the designed object by layers. When the process was finished, the final object was removed from the bed of "uncooked" powder, and any leftover powder shaken out.

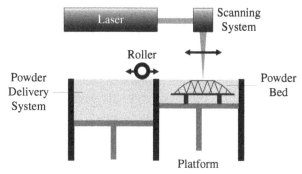

FIGURE 1.7 SLS 3D printing.

Deckard refined this technique to earn his master's degree in 1986 and his doctorate in 1988, having filed and been issued a patent for his invention in 1987. Seeing the potential for his technique, he had formed a company, Nova Automation, in 1986, later rebranded to DTM. The firm continued to grow and develop new machines, and it was acquired by 3D Systems in 2001 for US$ 45 million. Deckard's technique was soon extended to metals, which led to a range of similar methodologies.

At the heart of all of them is the fusing of materials to form the layers of the intended object. As with Deckard's manual approach, raw powdered materials are spread across a plate using a blade or a roller, fed by a reservoir located below or alongside the bed. The surface of that layer is then exposed to a high-energy beam that melts or "sinters" the material, after which the plate is lowered and another layer of material spread, before repeating the sintering process.

Today, companies like 3D Systems and Sintratec use lasers to melt plastic polymers using Selective Laser Sintering (SLS) (Figure 1.7). Others, such as Germany's SLM Solutions Group AG and the British firm Renishaw, use metal powders with Selective Laser Melting (SLM) and Direct Metal Laser Sintering (DMLS). The Swedish firm Arcam developed Electron Beam Melting (EBM), which directs beams of electrons over metallic powder layers using electromagnetic coils, the whole process taking place in a vacuum. Another variation, Selective Heat Sintering (SHS), is slightly different from the other variations in that a heated thermal print head fuses the powdered material together.

Following a much-heralded announcement that it was to enter the 3D printing arena, the technology giant HP unveiled its first family of 3D printing machines in 2016, and these were based on Multi Jet Fusion (MJF). This variation heats a layer of polymer material uniformly before jetting a fusing agent—microscopic drops of infrared-sensitive liquid materials—over the surface in the required shape. A subsequent detailing agent is then targeted at the contours to improve resolution. Finally, a lamp is passed over the surface to prepare it for the next application of material. This approach uses the many years of knowledge that HP has acquired in 2D printing to the point of using similar equipment, as pointed out by Paul Benning, Chief Technologist at HP:

> "The print heads are the same as those we use in our industrial inkjet printers."[6]

Today, the range of materials available for PBF is increasingly broad: thermoplastic polymers, metals, and alloys (including stainless steel, aluminum, titanium, and cobalt chrome), and glass and ceramic powders are all available, making PBF the preferred option for end-use products. New developments in conductive materials are also on the horizon, building on their already clear advantage: the "unfused" materials can act as supports if the designs have overhangs, negating the need for extraneous additions. These techniques can produce very-high-specification objects with great levels of precision. EBM, for example, is used in the aerospace and dentistry sectors when designs have complex geometries, thin-walled structures, and internal hidden channels and voids. The quality of finish and the tolerances that are achievable are limited by the fineness of the materials employed, which itself drives the costs of those. Furthermore, the size, shape, and homogeneity of the materials affects the quality of the fabricated parts.

One drawback of PBF is that surplus material within structures needs to be removed on completion, something that is not easy when dealing with complex forms. If it is collected and processed properly, though, this extra material can be reused, reducing wastage. The speed of the actual fabrication is also a constraint, with PBF techniques being relatively slow. Moreover, there are size limitations to what can be made due to the scale of the machinery needed. Of course, as high-energy systems are integral to the equipment used, these have a high power consumption. Furthermore, the combination of lasers, gas chambers, and an assortment of other systems mean that PBF-based 3D printers are expensive items.

Directed Energy Deposition

Directed Energy Deposition (DED) (Figure 1.8) builds on the material extrusion technique. As with those, material is deposited on a surface by a nozzle, either as a wire or in powdered form, the former being less precise than the latter,

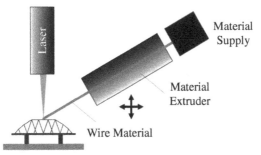

FIGURE 1.8   DED 3D printing with wire.

although more efficient in terms of material wastage. Unlike those extrusion techniques, though, with DED the nozzle can move along several axes, placing the material from any angle where it is melted. The most common variations are Sandia National Laboratories' Laser Engineering Net Shaping (LENS), US firm Sciaky's Electron Beam Additive Manufacturing (EBAM) and Wide Arc Additive Manufacturing (WAAM). They are similar to some of the FDM techniques, with two big differences: the material is only placed where is it needed rather than across an entire layer, and a Computer Numerical Control (CNC) machine can then be used to achieve an optimal surface finish with accurate dimensions. WAAM has the added advantage that it can be used to fabricate large metal items with low costs.

With these degrees of freedom, DED is typically used to repair items or to maintain parts that are integral to some finished item, using metallic raw materials such as titanium, cobalt, or chrome. The ability to control how the materials are laid down means that DED can be used to make high-quality parts, although naturally there will be an inverse correlation between the quality of finish and the speed of the equipment.

Sheet Lamination

As their name implies, Sheet Lamination methods use sheets or ribbons of materials to build up items. Each layer of material is placed on the bed and bonded to the previous layer. In the case of Ultrasonic Additive Manufacturing (UAM), the materials are usually metallic, and these are bonded using ultrasonic welding. With Laminated Object Manufacturing (LOM), which was developed in the early 1990s and used to create prototypes for the automotive industry, these are replaced with paper and glue (Figure 1.9). Once bonded, the top layer is cut by laser to produce the required geometry before the next layer is then added,

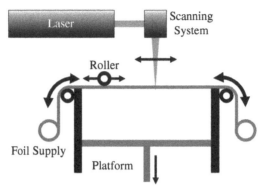

FIGURE 1.9 LOM 3D printing.

although some printer manufacturers reverse that order, cutting the material first and then bonding it.

A more recent development based on LOM is Selective Deposition Lamination (SDL) which produces paper objects by applying adhesive on top of the first manually-placed sheet of paper. The adhesive is applied selectively to define the layer of the paper, with the areas of higher adhesive density becoming the item to be made, while those areas with a lower density of adhesive serving as the support (Figure 1.10).

One immediate advantage of these techniques is that they work at low temperatures, reducing their energy consumption. Moreover, the materials used are inexpensive, reducing the cost of the process. The limitations in the type

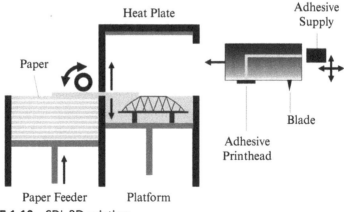

FIGURE 1.10 SDL 3D printing.

of materials—itself quite a small set—mean that the integrity of what is made depends on the adhesive used. Further, the finish can be quite variable, so post-processing is needed to finish products.

THE 3D PRINTING PROCESS

Whichever technique is employed to make things, the 3D printing process involves similar stages (Figure 1.11). The scale and duration of each one will depend on factors such as what is being made and how the object is to be used, with what dimensions, in what material, and using which technique.

The process begins with *design*, which usually involves the use of Computer-Aided Design (CAD) to draw what is needed. The starting point may be a clean sheet or a previously used model from which to develop the final design. Alternatively, designs can begin from reverse engineering, with the form of the object being captured using 3D laser scanners, magnetic resonance imaging (MRI), or other scanning equipment. For example, MRI scans have been used to produce the design of internal organs and bones of patients. The CAD packages that are used for 3D printing need to be set up with the limitations and parameters of the technology that will be used, such as tolerances, thicknesses, and geometries. Moreover, the designer needs to take into account how the object will be made in the 3D printer and consider whether supports should be incorporated into the design, even if only for the period of production.

Once the design is composed, it needs to be *modeled* and *simulated* in the conditions under which it will be made to check its internal strengths, stresses, and fatigue profiles, and to examine how it will behave under the conditions where it will be employed. For example, simulation software will look at the 3D printing process itself, analyzing how the materials will behave when they are melted by laser or electron beams, or how the heat will dissipate from an

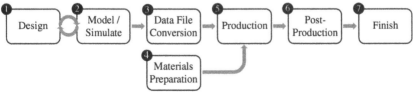

FIGURE 1.11   Typical 3D printing end-to-end process.

object and affect the residual internal stresses. It will look at how metal powders behave to form larger structures, and analyze the finished pieces, looking at, for instance, whether a 3D-printed engine mount will survive the air flows, torques, and strains it will have to endure on the wing of an aircraft throughout its lifetime, or fail catastrophically. These tools will include Computer Aided Engineering (CAE) packages that carry out the design analyses to identify flaws and potential issues that might arise during the manufacturing, postproduction, and operational processes. The analyses are computationally intensive and use many datasets upon which to base their conclusions. The outcomes are then fed back to the designers who tweak their designs, refining them iteratively until they arrive at a satisfactory version. The data file is then *converted* into the right format for 3D printing. Typically, the design is rendered into slices that will be fed to the 3D printer in turn (Figure 1.12). Additionally, Computer Aided Manufacture (CAM) tools are loaded with the necessary parameters to then control the various production tools.

In parallel with preparing the designs and data, once the 3D printing technology and necessary materials have been identified, the raw materials first must be *prepared*, then loaded into the 3D printer. Some materials need pretreatment to remove humidity, for instance, or to ensure that larger grains of powder haven't all traveled to the tops of containers (the so-called muesli effect, where larger grains travel to the top of containers of mixed grain sizes when the entire container vibrates while moving).

Next, the 3D printer is itself primed and prepared, and then *production* begins. If the fabricated object is attached to a printing bed or plate, or sits in a container of material or a vat of resin, it will need to be removed. Supports may likewise need to be removed, and both these activities may require machining, using cutters or manual intervention. Extra material, such as

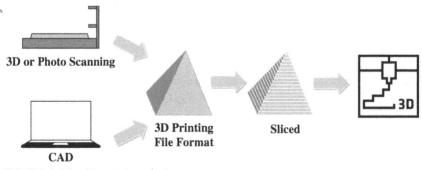

FIGURE 1.12   3D printing design stages.

unfused powders or liquid photopolymers and resins, may need removal from the internals of the item. Many 3D printing techniques then require some form of *postprocessing*, such as UV curing, or heating metal objects to relieve stresses that are generated during production. Lastly, objects may require some level of *finishing*, involving mechanical methods such as sanding or CNC machining to provide the tolerances and quality of finish that are needed, or the application of protective layers, paints, and coatings. Postprocessing and finishing should not be underestimated: they can easily be responsible for more than 75% of the total work needed to make a fully ready, quality item using 3D printing.

Compared to the situation 15 years ago, the 3D printing world is now far more capable and diversified, and developments continue apace. Although each technology brings with it different capabilities, there are common aspects of 3D printing that need to be understood if supply chains are to entertain the idea of employing it. That includes what 3D printing offers and the constraints that it has, which will be examined in Chapter 2.

Notes

1. H. Kodama, "A Scheme for Three-Dimensional Display by Automatic Fabrication of Three-Dimensional Models," *IEICE Transactions on Electronics,* J64-C, no. 4 (1981): 237–41.
2. Pagan Kennedy, "Who Made That 3D Printer?" *New York Times,* November 22, 2013.
3. John Hornick, Rob Wells, and Simon Lu, "3D Printing Patent Landscape," *3dprint.com,* July 17, 2017, https://3dprint.com/181207/3d-printing-patent-landscape/.
4. Aaron Marsh, "Three Ways 3D Printing Is (Quietly) Revolutionizing Trucking," *FleetOwner,* August 2, 2016, http://fleetowner.com/technology/three-ways-3d-printing-quietly-revolutionizing-trucking.
5. "Selective Laser Sintering, Birth of an Industry," University of Texas at Austin, Cockrell School of Engineering, Mechanical Engineering, December 6, 2012, http://www.me.utexas.edu/news/news/selective-laser-sintering-birth-of-an-industry.
6. M. Mitchell Waldrop, "3-D Printing Finds a Custom Foothold in Manufacturing," *Knowable Magazine,* May 2, 2018, http://www.knowablemagazine.org/article/technology/2018/3-d-printing-find-custom-foothold-manufacturing.

CHAPTER TWO

What Can and Can't 3D Printing Do?

T HE EVER-EXPANDING list of stories about new uses to which 3D printing has been put would make it seem that it can do pretty much everything. Without a doubt, 3D printing can produce things that were simply not achievable even a few years ago, making it very useful to supply chains in many ways. Its benefits are typically summarized in terms of the freedom it brings to design, the speed of setting up manufacturing, and the ability to make things where they are needed. The reality of 3D printing often astounds those who aren't familiar with the latest developments.

It is very easy to see only the positive—the continuous sequence of new capabilities and new uses—and it is just that optimistic view that leads to the overblown claims and assurances that many exponents of the technology make. Given how long 3D printing has been around, and the numerous positives that it brings, one is prompted to ask: Why isn't 3D printing already widespread? If it is so wonderful, why is it not being used to make everything that is manufactured today? For instance, 3D printing now represents about 1% of the US$ 120 billion global machine tool market—why is that figure not higher? To answer these questions, we must modulate the positive with the negative, separating reality from the hype.

Wonderful as it may be, 3D printing is centrally focused on making things using equipment, so naturally there are some needs that won't change: setting up the machines in the first place, having technicians to clean and tweak the machines, regularly replacing moving parts, and so on. Beyond this, however, are several limitations specific to 3D printing that inhibit its wider adoption. Many of those limitations are being pushed back, partly driven by the expectations that many have for it. As Richard Hague, Professor of Innovative Manufacturing at Nottingham University said in 2017:

> "Much of the hype surrounding 3D printing has also played an important role in advancing the technology."[1]

Understanding the capabilities and limitations of 3D printing is critical to answering whether it is suitable for supply chains and determining how best to employ it. This chapter will take a closer look at both factors, specifically addressing the following aspects:

- The *speeds* of 3D printers
- The range of *materials* that can be used to 3D print things
- The *designs* that can be realized through 3D printing
- The *design tools* that are needed to 3D print things
- The *volumes* of items that can be produced with 3D printing
- The *finish* of 3D printed items
- The *cost* to 3D print things
- The *locations* where 3D printing can take place
- The *labor* requirements involved in 3D printing
- The *sustainability* implications of 3D printing

 SPEED

Whichever technology is used, 3D printing an object is a slow and repetitive process compared to traditional reductive manufacturing techniques, including modern ones like Computer Numerical Control (CNC) machining. Even the simplest of objects can take hours to produce, as the 3D printer builds them layer by layer or dot by dot. Complex, high-quality items and those that use exotic materials can take days to finish. For example, Fusion Deposition Modelling (FDM) technology achieves speeds of between 50 and 150 mm/hr, and stereolithography (SLA) proceeds at 14 mm/hr. Other factors add to that time:

calibrating printers, preparing materials, programming design software with the right settings, and transferring digital designs between computers and the printers themselves. However, this does not tell the whole story.

Manufacturing something in the traditional method involves setting up and calibrating the production equipment before making things. Those items typically need to travel through production while being worked on using a range of reductive techniques. Once they're completed, the machinery needs to be cleaned or recalibrated ahead of the next run, a process that can be lengthy if a different configuration of design is required. 3D printing doesn't have those long stages. The printers are usually smaller, and faster to set up and calibrate than traditional manufacturing equipment, and the cumulative time saved by using 3D printing can be significant. Once it is set up, there is little need to reset it to accommodate differences in designs, further saving time. As with traditionally manufactured items, 3D-printed items typically require postproduction finishing, which does extend the total cycle time. Even taking this into consideration, 3D printing is significantly faster end to end than traditional manufacture, and this advantage makes it particularly useful for many supply chains both in making prototypes and producing end-use items.

For example, guitar manufacturer Fender used 3D printing to reduce the time it took to develop and introduce its G-DEC® amplifier. This product underwent various design and market research iterations to arrive at its final form, something that was achieved in under 6 months; using traditional techniques would have taken between 6 and 12 months for the design stages alone. Part of that acceleration stemmed from reducing the prototyping time from one or two weeks to two days.[2] Aerospace firm SpaceX cut the time to manufacture valve bodies from months to two days by switching from casting to 3D printing them,[3] and NASA has publicly stated a target of reducing the time to produce rocket engine parts 10-fold by replacing traditional manufacturing processes with 3D printing. In another example, a collaboration led by the Port of Rotterdam produced a 3D-printed propeller for a tugboat, reducing the production time for such an item from several weeks to just under 11 days[4] (see sidebar).

This faster time to manufacture is opening new possibilities in sectors that usually don't make things. The Canadian company C3DE is using 3D printing to create what it calls "compelling evidence," items "to help lawyers, paralegals and expert witnesses bolster their case activities." Problems arising from credibility and memory bias that often plague legal proceedings can be solved using accurate visual aids. By using models that are rendered quickly

Propelling 3D Printing

The Port of Rotterdam is the busiest shipping port in the world, handling 30,000 sea-going ships and 105,000 inland vessels every year. With so much traffic, many ships use the port as a repair stop. One critical item that often requires replacement are ship's propellers, which get damaged during operations. Replacement can be a long, convoluted, and expensive process that requires dry docking, during which the ship is not earning. Finding ways to shorten that timeline therefore becomes imperative. Being large, precision-manufactured metal objects, propellers have long lead times. If Rotterdam shortens those lead times, it has to keep some in storage—an expensive prospect given their size and weight.

In 2016, the Port came together with two technology partners, InnovationQuarter and RDM Makerspace, to form RAMLAB. The aim of this venture from the outset was to produce certified metal parts on demand using 3D printing. Together with Damen Shipyard Group, Promarin, Autodesk and Bureau Veritas, each bringing a particular expertise, RAMLAB focused on one of the items with the longest lead times to produce a ship's propeller. The final design was for a 1,350 mm diameter, 200 kg object baptized WAAMpeller, chosen because of the 3D printing technology that was to be used, to be made in a nickel-aluminum-bronze alloy, the standard material for such items. The first prototype was delivered in mid-2017, and the lessons learned from that were then incorporated in the manufacture of a second WAAMpeller. This was delivered in late 2017, having taken 256 hours to produce, far less than the several weeks' lead time for a traditionally cast version. This second propeller, made from 298 layers of alloy, was then installed on a tug for testing, and subsequently given class approval in November of that year, allowing it to be installed on ships and used operationally.

From the opening of RAMLAB, it took only one year to develop, produce, test, and deliver the first class-certified 3D-printed propeller, a significant reduction in the time to mount such a manufacturing operation compared to the time to establish and produce a traditional manufacturing plant for the same items. The achievements of RAMLAB have taken it a long way to its mission of accelerating the adoption of 3D printing in the shipping value chain.

using 3D printing, experts find it easier to explain and communicate issues in a court case. In one example of a case that used 3D printing, a broken spine was made from x-ray data to demonstrate the severity of a person's injury.[5] The faster production times have also allowed surgeons to use 3D-printed models of patients' physiology to help practice for operations, with iterations made to understand the approach and effect of surgical procedures—impossible without the faster production cycles of 3D printing.

While it may be faster than traditional manufacturing, this isn't always the case, particularly when there is a need for mass manufacture. For instance, consider an item that is made using injection molding where thousands of objects can be made very quickly; some machines using this approach can produce 144 plastic forks with a cycle time of 9.5 seconds.[6] However, 3D printing has been found to help here, too. Supply chains that rely on injection molding are increasingly leveraging the swifter design phases, particularly for 3D printing the molds, by reducing the time to perfect and produce them. This has allowed those supply chains to respond more rapidly to changing customer requirements and competitive pressures.

 MATERIALS

From its inception, 3D printing has been a plastics-based technology—predominantly white plastic at that—and it wasn't until the turn of the twenty-first century that the range of materials expanded significantly. New techniques, such as FDM and Selective Laser Sintering (SLS), increased the number of materials that could be 3D-printed while customers clamored for more choice in quality and variety. Today, not only are the companies that make 3D printers investing in R&D for different technologies, so too are the companies that make the raw materials as well as those that use them, such as GE, Boeing, and Nissan, all of which are investing significantly to develop new metal powders, better plastics, and a wider range of other materials. For instance, the leading light-metals producer, Alcoa, opened a dedicated US$ 60 million production facility in early 2016 specifically aimed at producing proprietary titanium, nickel, and aluminum powders optimized for 3D-printed aerospace parts.[7] The pursuit of better materials has also affected university departments, where there has been an uptick in the number of materials science courses and postgraduate research projects devoted to this field, largely sponsored by companies that seek to improve the variety, applicability, quality, and cost of those materials. Although the range of materials available to traditional manufacturing processes will continue to outstrip what is available for 3D printing for some time, new materials in new grades are being announced monthly.

Today's 3D printers can produce objects in a wide number of materials:

- *Plastics.* The progress in plastic materials science over the last 30 years means that objects can now be made with the particular characteristics needed for how they will be used and the conditions in which they will be employed. This includes materials with different levels of malleability, in a wide variety of colors and textures, and suitable for use in high- and low-temperature environments. Today, the most common plastics used for 3D printing include polylactic acid, acrylonitrile butadiene styrene (ABS), and polyamides (such as nylon). Other materials include high-density polyethylene, high-impact polystyrene, and polyvinyl alcohol, which tend to be found in more everyday objects in traditional manufacture. Plastic-based 3D printers are generally SLA and FDM-based, as the characteristics of the materials, including their low melting points, make them well suited to those techniques.
- *Metals.* From an industrial viewpoint, the developments in this area of 3D printing are the most significant. Although metal powders remain a small part of the metals market, 3D printers can now make items in bulk metals, such as iron and copper; harder metals and alloys such as titanium, stainless steel, and Inconel; and precious metals, like platinum, gold, and silver. Specialist firms specifically aimed at the 3D printing metal materials market are now emerging. The UK-based company Metalysis processes titanium ore using electrolysis to produce a high-grade titanium powder for use in 3D printing. Welding titanium metal parts is difficult and expensive, but 3D printing them offers all the benefits of 3D printing in a way that overcomes those challenges. Metal 3D printing typically uses powder bed and Directed Energy Deposition–based techniques, like SLS and Electron Beam Additive Manufacturing (EBAM). They are increasingly capable of rendering finished industrial products, although some level of reductive postprocessing, usually CNC, is required. The key to which metals and alloys can be used lies in how they act when in molten and powdered forms for use as 3D printing raw materials, and in their characteristics during and after manufacture; this is where much of the current research and investment in 3D printing materials is focused.
- *Organics.* New techniques based on material jetting, SLA, and FDM are now able to produce objects from organic matter, including wood resin and pulp, foodstuffs, and, most intriguingly, organic tissues, usually in liquid or suspension form. For instance, the company Organovo made blood vessels from a donor's cells in 2010 using bioprinting—essentially 3D printing

using organic matter termed bio-inks—and in October 2016 started to offer liver-cell-based products for use in pharmaceutical research. The company, founded in 2007 in Delaware to advance bioprinting technologies, has an ambition to 3D-print heart, lung, and kidney tissues, and in 2015 they partnered with the global beauty firm L'Oréal to supply 3D-printed skin, removing the need for animal testing of cosmetics.[8] This capability promises to revolutionize medical treatments, and institutions including the US Department of Defense and Wake Forest University are investigating the use of 3D-printed organic material to treat wounds and burns.[9] The Swedish company Cellink uses 3D bioprinting to build cartilage and skin cell-based structures, for use predominantly in drug testing and cosmetic sectors,[10] and already they can create anatomically precise body parts such as noses and ears. Reza Sadeghi, chief strategy officer at Biovia group, part of Dassault Systèmes, expects the first such organs to be used in live reconstructive surgery by the early 2020s.[11] Given that in the period 2007–2017, 6,000 out of 49,000 patients needing organ transplants have died while waiting for donors,[12] it is easy to see why 3D bioprinting of replacement organs is a fast-developing branch of 3D printing.

This diversification of materials has been matched by developments in printer technology: printers supporting simultaneous multicolor production are already available, while others are starting to produce parts in multiple *materials* in one go, significantly reducing the time to produce a finished object or component. Whereas before, parts would have to be made in a single material, then processed again with others, then assembled, now the number of stages can be reduced. This promises to be a step change in the adoption of the technology for industrial purposes, particularly when, for instance, sensors and electronic circuitry can be embedded within an object during printing. The industry is on the cusp of making this a reality, with new techniques and cutting-edge materials, such as carbon nanotubes and graphene, promising to reach that goal. Already, the company Voxel8 offers materials that use a proprietary conductive ink and printers that allow for a build volume of $10 \times 15 \times 10$ cm with a layer resolution of 400 µm; this opens up the potential for printing electronic items. The company is already developing ink designs for producing resistors, lithium ion batteries, and stretchable electronics and sensors.[13]

It is worth noting that at various times, some 3D printer makers have sought to restrict the materials that can be used on their machines, ostensibly to ensure their smooth operation and the quality of what is produced. Of course, from the perspective of the printer makers, this is an attractive proposition, akin to the "genuine ink cartridges only" restriction that some 2D printers have had in the past, which allowed only "legitimate," branded cartridges to be used with them rather than cheaper—and possibly damaging—ones. In 2017, HP printers began to reject ink cartridges produced or refilled by third parties, something that they had to later compensate customers for. It was the market that put pressure on HP and the other 2D printer makers to remove those restrictions, and likewise it will be the market that will demand openness from 3D printer makers regarding the materials that can be used with a particular machine.

Questions of how and where final production items are to be used inform the necessary characteristics of the materials needed to make them. This includes understanding the stresses, strains, and torques that an end product will be subjected to, all of which affect the required characteristics of the materials that parts are made of. When making models and prototypes out of plastic, those characteristics aren't particularly important, so long as the plastic holds its integrity for as long as is needed. In many areas, though, those characteristics are vital, such as in safety-critical parts and systems, and this is as much the case for 3D-printed parts as it is for those that are traditionally manufactured. As 3D printing has moved into manufacturing finished items, questions of strength, malleability, and integrity have become more important. Certainly, in the early days of metals 3D printing, there was a suspicion that the quality of items made was not adequate, and those concerns—though attenuated over time through R&D and investment—continue. In 2013, NASA conducted a series of tests on the metallurgy of 3D-printed fuel injection nozzles designed for use in its rocket engines. Those tests revealed that the integrity of the 3D-printed parts, made using FDM, was actually *better* than traditionally cast parts, and the nozzles were perfectly capable of handling the high temperatures and pressures in an operating rocket engine.[14] Indeed, these outcomes have been found in many other analyzed items in other sectors, from energy to aerospace and defense.

There is good cause for caution: researchers at Carnegie Mellon University studied some titanium parts by carrying out intense tests on them with X-ray synchrotrons and microtomography. What they found was that parts made of one commonly used titanium-aluminum alloy powder, Ti-6Al-4V, using SLS and Electron Beam Melting (EBM) 3D printers, contained very small bubbles of gas within their finished structures. Those bubbles tend to act as pinpoints of weakness, reducing the overall strength of the finished object.[15]

The researchers established that they stemmed from the manufacturing method itself: when the powders are melted, gas is trapped in the liquid metal, resulting in the higher porosity in the cooled, solid form. That porosity acts as a catalyst in the weakening of the metal object, accelerating cracks and faults. However, the study also found that this issue can be ameliorated, if not mitigated entirely, by adjusting the power, speed, and placement of the printing equipment. These are still the early days of such analyses, and more investigation is needed; the results will inform the evolution of 3D printers and materials design.

Considering wider materials issues, those used in 3D printing naturally face the same concerns regarding their sources as do those used in traditional manufacture. Plastics and resins are made using oil derived from plants and petroleum; metal powders are developed from ores. In all cases, the same concerns of sustainability, environmental protection, and human exploitation apply. For instance, several ores are sourced from conflict areas, such as the Democratic Republic of the Congo. Efforts are underway in various university departments and start-ups to extend the use of recycled materials in 3D printing, a topic to which we will return later in this chapter.

DESIGN

When things are made using reductive methods, the designer needs to consider how it will be made, in terms of the materials, equipment, and processes to be employed. Traditionally, that means adjusting designs to accommodate the needs for drilling, milling, filing, and shaping—all the ways of removing matter to produce the desired object. For example, the design must allow for how a lathe works or for a drill bit to access the right spot for boring into the material. With 3D printing, those restrictions are lifted: in effect, if the form of an object can be visualized, it can be produced, with few or no changes required. 3D printing allows for complexity in designs that would otherwise not be possible using traditional manufacturing techniques by virtue of its approach of building items layer by layer. This greater freedom allows for design parameters that would be unachievable using traditional techniques, such as reducing the thickness of internal structures or employing porous internals rather than solids.

With this freer complexity, designs can now have increased functionality or be wholly optimized for a specific need without compromising the manufacturing process. As John Jaddou, cofounder of the design consultancy Addeation says:[16]

> "With traditional manufacturing, we're limited to certain geometries and materials, so we must design for the process—whether it's injection molding or casting, for example—but doing that constrains design freedom. With [3D printing], we have much more flexibility and the ability to leverage a broader engineering tool chest, like topology optimization, which allows us to optimize the design for functionality."

This means, for instance, that items can be created that are optimized to have the lowest weight while retaining structural strength, or for variable stiffness or for complex fluid flow. In 2013, GE set an open challenge to the manufacturing world: to redesign an aerospace engine bracket with the goal of reducing its weight while retaining the necessary strengths, using 3D printing as the means of production. By 2016, the most successful submissions, proposed by a collaboration between 3D Systems and the software firm Frustum, saw a 70% reduction in weight while meeting every single other requirement. Rather than being "chunky" in form—and thus easier to manufacture using traditional techniques—the new design has a complex, organic-like shape (see Figure 2.1), something that would be challenging to manufacture with reductive methods but perfectly suited for production using 3D printing.[17] The benefits from that redesign come from reducing aircraft loads (and thus fuel costs), engine emissions, and material used to make the bracket.

Such topology optimization has been used by the high-end automotive manufacturer Bugatti, part of the German Volkswagen Group. In 2018, the

FIGURE 2.1   Transformation of design before and after optimization.
Source: Image courtesy of Frustum/3D System.

company unveiled a single-piece titanium brake caliper made from metal powders that were aimed at producing parts destined for high-stress situations in the aerospace sector, such as undercarriage and wings. It is made using 2,213 layers of material, taking 45 hours to produce, and it is then heat-treated in a furnace to release residual stresses before it is removed from the printing tray and finished using a series of reductive techniques. At 2.9 kg, the resultant caliper is 2 kg lighter that its machined aluminum predecessor, while at least matching the strengths that it must endure. Constantly on the lookout for other improvements, the firm has also produced a 63 cm long, lightweight aluminum windshield wiper arm weighing just 400 g.[18] At the other end of the size scale, the drug maker Aprecia Pharmaceuticals has developed a pill for its drug Spritam, used to treat epilepsy, that is more porous than previously possible because of 3D printing. This feature makes them easier to dissolve in liquid, and thus easier for patients to swallow, particularly for higher doses.[19]

Another aspect of enabling freer complexity is—ironically—that 3D printing allows products to be simplified. Traditionally, items are made in several parts that are then assembled, usually because of manufacturing constraints or the need for different materials to cope with particular manufacturing conditions. This results in the need to create highly accurate fits between the surfaces of the items to be assembled, adding to the manufacturing time and effort needed. It also leads to weaknesses in those items, with a risk of damage and failure at the joins of the assembled components. 3D printing allows entire objects to be made with fewer parts or even as a whole with no assembly required, termed "holistic manufacture," which reduces or eliminates those drawbacks. GE has led the way here in the commercial arena with a well-publicized example: fuel nozzles for its next-generation Leading Edge Aviation Propulsion (LEAP) engines (Figure 2.2), which are fitted on the new Airbus A320 Neo and the new Boeing 737 MAX aircraft.

The nozzles for those engines inject fuel into them and operate at very high temperatures—typically 1,650 °C (3,000 °F)—and pressures.[20] They therefore need to be durable and reliable, with precise tolerances to ensure an exact delivery volume and rate of fuels. The traditional nozzle, assembled from some 20 separate parts, had a wildly complex internal arrangement. Each of those separate parts would be sourced from different internal or external suppliers and then painstakingly welded and brazed together. When the LEAP engine design team reviewed this component, they decided that 3D printing could simplify the item itself and provide significant benefits in time to

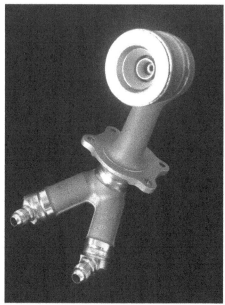

FIGURE 2.2   GE's LEAP Engine Fuel Nozzle.
Source: Image courtesy of GE.

manufacture, as well as finished weight and cost of production. Consequently, they set about redesigning the nozzle. They chose to use SLS/Selective Laser Melting (SLM) techniques to resolve the challenges of the internal structure. Each nozzle is then CNC-machined to produce the necessary high precision and quality of finish. The end result is a single, 3D-printed, extremely precise metal fuel nozzle that not only has a durability and strength that surpass those of its predecessor fivefold, but that is also 25% lighter. Its efficiency is higher, improving fuel burn by 20% and providing 10% more power to the jet.[21] The economic advantages are undeniable: when the reduction in weight is multiplied by the number of nozzles in an engine and the hours of operation that an engine typically accrues, this represents an annual fuel saving of about US$ 3 million per aircraft. By the end of 2017, GE had taken orders for more than 12,200 LEAP engines—an order book worth some US$ 170 billion at list prices—with each engine having 19 nozzles.[22] Moreover, the reduction in production time due to there being no need to assemble and weld the item means that the manufacturing cycle time is far reduced compared to traditional manufacture. Such is the success of this single component that GE is rolling out dedicated 3D-printing factories, first in Auburn, Alabama, with 40 machines, and then with a US$ 200 million manufacturing plant in India. These will produce a wider range of 3D-printed products across several GE divisions, from aerospace and transport to power distribution and oil and gas.[23] The company is ramping up fuel nozzle manufacture from their current production level of 14,000 per annum with a target of reaching 40,000 by 2020. More widely, GE expects to use a similar approach to make other parts of the LEAP engine with 3D printing, with a view to reducing the weight of an engine from 6,000 to 5,000 lb. In 2017, Mohammad Ehteshami, the former Head of Engineering at GE Aviation said:

> "In the design of jet engines, complexity used to be expensive. But additive allows you to get sophisticated and reduces costs at the same time. This is an engineer's dream. I never imagined that this would be possible."[24]

The other major jet engine manufacturers have similarly embraced 3D printing with similar results; Pratt & Whitney is using it to make fasteners, fuel collectors and, like GE, fuel injection nozzles in nickel and titanium with EBM and Direct Metal Laser Sintering (DMLS) technologies.[25] The use of 3D printing has led to a reduction of around 15 months in their whole engine design process, with a 50% saving in the final weight of some of those parts. Rolls-Royce has been using 3D printing for over a decade to help in the design and development of new engine components, and in 2015, the company, together with the UK's Manufacturing Technology Centre in Coventry, UK, the University of Sheffield, and the 3D printing technology firm Arcam, made the largest 3D-printed component for an aircraft engine, a 1.5 m × 0.5 m titanium front-bearing housing using EBM.[26]

It was these sorts of simplification advantages that encouraged researchers at the Oak Ridge National Laboratory to redesign a robot arm for the US Army. Typically, the hydraulics pipes that drive such arms are separate components, external to the arm itself. The research team simplified the design by incorporating the pipes as ducts within the structural parts of the arm, which reduced the parts list from 238 items to 28.[27] The new design is only possible through the use of 3D printing manufacture. Naturally, reducing the number of components and assembly steps also brings cost savings. When NASA redesigned the environmental control duct assembly system for the Atlas V rockets, reducing it from 140 metal parts to 16 using a Stratasys Fortus 900mc 3D printing production system and ULTEM 9085 FDM thermoplastic materials, it cut the cost of making the item by 57%.[28]

The significance of all these changes is that a new design paradigm is required, one that is unconstrained by the restrictions of traditional manufacturing and that considers the freedoms and limitations of 3D printing. Designers will now consider producing things in fewer parts or singly rather than assemblies of components, but they will also be cognizant of being too ambitious. This is a key consideration in reducing supply chain complexity, one we will examine more closely later in this book.

Despite their advantages, complex designs have their own constraints in manufacture, finishing, and testing. For instance, gravity still applies, and when making some designs, the manufactured objects may need external supports in the 3D printer that can be detached afterward, usually through traditional or manual methods—complexity, it turns out, is free-ish! The more complex the design, the more is needed from the Computer-Aided Design (CAD) software used to develop it, and that capability is not infinite. There comes a point when complexity is such that humans cannot produce the design—imagine trying to design the internal structure of a human liver, with its minuscule tubes and ducts numbering in the hundreds of thousands. To cope with that, there will be a need for more automated design using artificial intelligence, and research is already underway to achieve that. Those complex lattices and structures all need to be simulated to test the designs and verify their fitness for purpose. This includes checking that the design is printable, verifying thickness of walls and gauges of ducts, and so on. Such tests are typically achieved with specialist software that also needs to be included in the development costs.

Complexity can also be a function the accuracy, precision, and tolerance required in a part, and 3D printing is increasingly able to make items more precisely and accurately than traditional techniques alone have previously allowed, particularly when it is used in conjunction with the most modern techniques such as CNC machining. For example, dentists employ 3D printing to build crowns and implants, either directly or made from casts, and the tolerances involved can be of the order of micrometers. Inserting hydraulic pathways in an object, however, requires high precision and accuracy, and unlike dental prosthetics, is something that otherwise would not be possible using reductive techniques. However, freedom of design doesn't come without consequences, and designers need to be cautious not to be overzealous: the more precise a design and the lower the required tolerances, the longer it will take to make in a 3D printer. Likewise, making structural elements with thinner walls or partitions increases the risk of buckling, reducing the strength of the item.

Accuracy, Precision, and Tolerance

These three terms are frequently—and erroneously—used interchangeably by non-manufacturing people, so it is worth clarifying what is meant by each one. To help identify the distinction, consider an archer firing arrows at a target.

- *Accuracy* refers to how close the result of a measurement, calculation, or specification conforms to the correct value or a standard. In the example, the bull's-eye is the target. How accurate a shot is depends on how close to the bull's-eye it hits. In the case of manufacturing, the CAD design is the bull's-eye, the correct value that the printer must achieve.
- *Precision* refers to the exactitude of a measurement and hence its repeatability. In the example, the archer is precise if their shots hit the same point every time, even if that point isn't the intended one, such as the bull's-eye. In the case of manufacturing, this refers to the reliability with which the machinery makes things the same way time and time again.
- *Tolerance* refers to the level of acceptability of accuracy. If the archer's arrows hit within an inch of the center of the bull's-eye, and this is acceptable, this is tolerable. If hits within than a half-inch are needed, then shots within an inch but still outside that tighter radius are out of tolerance. In a manufacturing sense, tolerance refers to the level of difference from the intended design that is acceptable.

Figure 2.3 illustrates the different combinations of accuracy and precision.

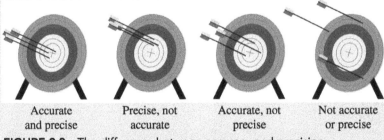

| Accurate and precise | Precise, not accurate | Accurate, not precise | Not accurate or precise |

FIGURE 2.3   The difference between accuracy and precision.

(continued)

(continued)

How does this translate to 3D printing? All three have a direct consequence on technology required, the cost of the 3D printing equipment, and manufacturing time. For instance, a prototype made for educational purposes doesn't need the same accuracy and precision as, say, a part for a jet engine. A dental implant needs a very high level of accuracy and precision and a low tolerance. Printers using the STL format (discussed in the next section) will generally not be a precise as SLS.

▓ DESIGN TOOLS

The 3D printing process starts with designing the objects to be made. Under-standing what is involved in that task and what can be done with today's tools are necessary when considering whether to employ the technology in a supply chain. That information helps to answer questions of what it would take for 3D printing to make what those supply chains require and about the limitations that must be overcome. The many different tools available to designers and the many file formats for encoding the designs each have different capabilities, and if supply chains are to leverage 3D printing effectively, knowing what these are and choosing the right one is an important decision.

A 3D printer must be fed with design information to print an object. The first consideration involves the file format of the designs that will be 3D printed. While CAD packages have long been used to create and store 3D designs, not all are able to do so for the purposes of 3D printing. Even when they are, many of the suitable formats don't support the data needs of many modern techniques and machines. While several formats have been used in 3D printing over the years, four dominate the industry:

- STL
- OBJ
- AMF
- 3MF

The STL Format

For much of the history of 3D printing, designs intended for it have been written and stored as STL files, an easy-to-use data format originally developed in the 1980s by Chuck Hull for the very first 3D printers. "STL" refers to the technique

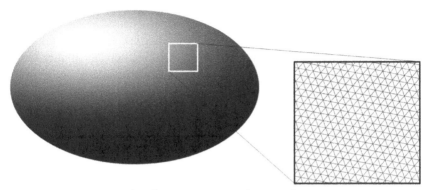

FIGURE 2.4   STL encodes shapes in a triangular tessellation pattern.

it was meant for and that he created: stereolithography. It allows for the encoding of a series of 3D shapes, or polyhedra, that describe the overall design using triangular tessellation as a basis, giving rise to the after-the-fact interpretation of the acronym of "Standard Tessellation [or Triangle] Language." Consider the example of an ellipsoid object, depicted in Figure 2.4. Once it is created in CAD software, the data file for the design is converted into STL which represents the surface as a series of tessellated triangles, whose spatial data is sent on to the 3D printer.

Given its age, STL is still a popular file format, with many software tools available that use it and a multitude of models readily available to be downloaded from sites such as Thingiverse and Shapeways.

However, given its vintage and the continual progress in 3D printing technologies since those early days, it is unsurprising that the capabilities of today's more advanced 3D printers surpass what the STL data format can offer. Modern printers and the objects they manufacture need more detail, and hence more data, to encode all of the information needed to print the target object. If one were to use STL for more modern purposes, such as printing objects with a very high level of precision, the data files would quickly become long, heavy, and unwieldy, often crashing the CAD system. If a very smooth surface is required, the triangles that are encoded can be made very small, but that expands the file size considerably. Alternatively, STL may not be able to convey the needed information; for instance, on materials, texture, and color, which weren't a problem when all that could be produced with a 3D printer were items made of single-color plastic. Furthermore, the STL format is prone to errors that are not picked up automatically and only appear when the objects are made, such as holes or overlapping triangles.

The OBJ Format

As new 3D printers emerged that could be used to make things in more than one color and material, a second format emerged as a preferred alternative: OBJ. This format was already in use by graphic designers, and the 3D printing community adopted it over similar formats because it is open source and relatively simple to use. The OBJ format encodes information on geometries like STL, representing them as tessellated triangles—thus carrying the same risk of having large files for smooth surfaces—or, in a more recent development of the format, as free-form curves and surface schemes, which reduce file sizes. Moreover, the OBJ format can store color and texture data in a companion Material Template Library (MTL) format and feed both files to a 3D printer. Because of this increased capability, OBJ and MTL have become the preferred choice of industry sectors requiring high levels of precision in their designs, such as aerospace and automotive.

Despite its advantages over STL, OBJ suffers from being difficult to "debug," to identify and resolve design issues. Moreover, the need for two files to represent an object means that configuration control can become troublesome very quickly, particularly in the industrial design situations that many companies encounter.

AMF and 3MF Formats

As data needs for 3D printing grew, driven by technological evolution, it became clear a new format was needed, one specifically aimed at more modern 3D printing. Many companies during the 1990s and 2000s sought to develop proprietary formats, typically tied to their own hardware. As with other technologies, they soon faced the push for more open source environments, so designers could use the same design data on as many manufacturers' hardware as possible. This, together with the need to ensure a standard that can support the long lifecycles of many 3D-printed designs, drove efforts to develop a new, common, international standard. The two leading contenders are AMF and 3MF.

Developed by an ASTM International task force, the AMF format arose in 2010 out of the recognition of the previous formats' shortcomings. This XML-based format allows for the specification of geometry, color, texture, and material in the design of an object, overcoming the restrictions of the older STL format. As an XML file, it is easier to write, read, and process, a significant advantage in itself. Rather than the flat triangles of STL, AMF uses curves, which cuts file sizes by reducing the number of shapes needed to describe them, and thus the amount of code. Moreover, it can store information on color, texture, material, and structure within the single file, overcoming the two-file drawback of OBJ. It also has a novel feature: allowing for multiple objects to be encoded in the same file and be 3D printed at the same time on the same print bed. ASTM ensured that the format was backward-compatible, so that if a 3D printer cannot produce items in many colors or materials, for instance, then the data on those in the design file is simply ignored.

Despite its greater versatility, the AMF format hasn't yet been widely adopted, particularly because the leading organizations in the 3D printing industry were largely not included in its development. As a result, CAD package developers and the 3D printer makers have each been waiting for the other to press ahead with its use.

In tandem with the creation of AMF, a Microsoft-led consortium formulated the 3MF format, something that Microsoft had been quietly including in its Windows 8 and 10 product development efforts.[29] The 3MF Consortium includes printer makers like 3D Systems, Stratasys, EOS, GE, HP and SLM Solutions, manufacturers like Siemens and Dassault Systèmes, software firms like Autodesk and nTopology, and 3D printing service platforms like Shapeways and Materialise, together with several associate members. This broad membership has given the 3MF format significant momentum to push for its wider adoption, beating out the ASTM-led alternative.

With echoes of the AMF format, 3MF provides a standard and open format that is more robust than previous file formats[30] and that is:

- Highly descriptive, able to include model geometry, color, and other characteristics;
- Adaptable, so that it can adapt to future requirements as well as 3D printing innovations; and
- Practical, easy to implement, open source, and interoperable between hardware

To achieve this, 3MF is also XML-based and uses triangular meshes, although in a manner that reduces file sizes compared to STL. Overcoming many of the shortcomings of past formats, 3MF includes a large measure of error correction, fixing some of the issues described under the other formats. As with AMF, it conveys data on color, texture, and material in a single file, together with a raft of further information to increase its functionality, all aimed at making 3D printing easier.

While the data formats are beginning to catch up with the technology and the needs of industrial supply chains, designing objects for 3D printing continues to be something best left to specialists. Even with those formats and software making the task easier, it is not particularly intuitive and even the simplest design software requires training and practice. Where more sophisticated and complex designs are called for, CAD, Computer Aided Manufacture (CAM), and Computer Aided Engineering (CAE) packages become tools for deep experts. While that restriction is clearly holding back the penetration of 3D printing in households—after all, we are not all CAD designers—it also affects industrial supply chains. Designers are often expensive bottlenecks, and good ones have become critical to employing 3D printing well. As most companies don't immediately have access to them within their own ranks—yet—they are usually contracted out to do the job via an external provider.

This hindrance to the accelerated adoption of 3D printing has been fortunately noticed by the CAD/CAM package providers, who are beginning to respond with more intuitive and user-friendly software. Some 3D printer makers now offer their own custom software, often as a service on the cloud. To assist designers and accelerate the iterative CAD/CAE processes, software designers are also increasingly looking to provide automation tools, which has the added benefit of reducing the risk of human error.

However, CAD/CAM software is only part of the software set that is needed. Many business cases for using 3D printing depend on optimizing designs, such as removing unneeded internal material to make parts lighter. Techniques for this include topology optimization, honeycombing, and using iterative processes to refine designs. All too often, the software packages that can do these tasks are separate from CAD/CAM, and there is a need to transfer datasets from one system to another in a controlled manner. While it may seem trivial, this slows down workflows and can lead to a breakdown in configuration control. Moreover, actually running these packages requires hours, sometimes days, for a single iteration to be ready. This issue has also been noted by software developers, who are beginning to unite the capabilities of CAD, CAM, and CAE into single suites, speeding up design cycle times.

Meanwhile, other solutions to this bottleneck have emerged. For instance, the last few years have seen the advent of specialist 3D-printing design houses, where experienced CAD designers can be hired per project. Companies such as ThinkSee3D in the UK take on the design role, from starting with blank sheets to creating 3D designs from laser or CAT scans of objects.

Another emerging trend is the use of open source, ready-made designs. A tangible example of the sharing society, individuals create designs for a wide variety of items and translate these into the requisite 3D-printing file formats, ready for use. The files are then uploaded to a cloud-based design repository where subscribers can access or amend them, either per item or as part of a term plan. Companies such as Shapeways, Thingiverse and Pinshape have already accumulated thousands of designs. Indeed, many supply chain managers already considering the technology have voiced that open source collaboration is a distinct area where 3D printing will play a part. In this situation, the digital designs can be shared with a growing number of open source partners, either companies or individuals. This will allow prototypes to be rapidly developed and tested, and those cycle times can be further accelerated by getting customers actively involved in those communities. Taking this concept further, mobile phone manufacturer Nokia opened the design specifications for its Lumia phones so that end users could design and produce cases for their phones, customizing them as they wished. One of the earliest companies to commercialize the flexibility of 3D printing was Freshfiber. Founded in 2009, they were the first to offer consumers a fully 3D printed, customized smartphone case, and today the goods on offer encompass a wide range of consumer products, jewelry designs, and homewares.

VOLUMES

Production level considerations and the costs of production shape supply chains, giving rise to minimum and economic order quantities (MOQ and EOQ). Scaling those comes at a price in the form of capital investment, and producing smaller quantities typically results in higher costs per item. 3D printing breaks this paradigm, allowing for production in lots of one—that is, a series of singly designed items. Using traditional manufacturing means time-consuming retooling of machinery, which interrupts production; 3D printing eliminates the need for this, allowing consecutive items to be made with little or no change to the machinery. This gives supply chains an opportunity to improve their levels of service and meet the specific needs of their current customers, while opening up their range of offerings and thus potentially expanding their customer base. It also is beneficial to start-ups by lowering the barriers of entry, enabling them to start making things quickly and more cheaply. This has the potential to completely transform entire industries. For example, it catalyzes the advent of personalized healthcare, where the physical elements of treatment, from implants to medical supports, can be customized to the individual patient at no more expense than having a standard, "one-size-fits-all" approach. According to Steve Tomlin, consultant pharmacist at Evelina London Children's Hospital, by having a 3D printer at the hospital, different shapes can be used for child medication, not only altering the dosage, but the size and shape of the tablets, to make them more personalized and even fun.[31]

This changes the historic rules that shaped and scaled supply chains, allowing companies to mass-customize items, from tools to fashion accessories. In a market where companies and consumers increasingly want personalized or bespoke items, this alone makes 3D printing an attractive proposition. However, this freedom is not complete: mass customization still requires the mass adjustment of designs, and each of those new variants needs to be tested or simulated. Those processes are time consuming and costly, requiring manual or low levels of automation, and the practicality and economics of mass personalization require careful analysis to justify them.

In its early days, 3D printing was predominantly used to produce single items, usually with low design complexity since the software and machinery weren't very sophisticated or capable (Figure 2.5). As the capabilities of both have advanced, so the complexity of what was designed and 3D printed increased. However, the higher costs of 3D printers and the materials they use, and the slower speeds of manufacture, all hindered its use in higher-volume production, particularly when compared to injection molding.

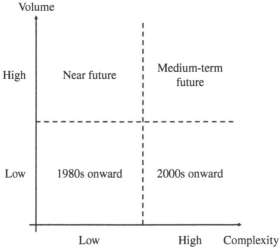

FIGURE 2.5   Development of 3D printing in terms of volume versus complexity.

It isn't all bleak, though. The first step in production is to acquire production equipment and set it to work, and the initial setup costs for traditional manufacturing are considerable, when equipment, facilities, and so on are taken into account. As production volumes increase, so the contribution of overheads to the cost per item drops, making higher volumes more attractive to produce, up to the maximum capacity of the machinery. 3D printing changes that dynamic. As the costs of 3D printing falls and the speed at which it makes things increases, the technology becomes viable at higher-volume manufacturing. By 2016, over half of US manufacturers was predicting that 3D printing would be used for high-volume production within three to five years, and that proportion is growing year to year.[32] With the advent of new operating models and mass 3D-printing facilities, that prediction is coming closer to realization. For example, Fast Radius (formerly CloudDDM) partnered with the logistics company UPS at their Worldport hub in Louisville, Kentucky, leveraging both companies' strengths to open an "end-of-runway" facility, and in 2016 they announced plans to open similar facilities in Singapore. This partnership brings an initial manufacturing bank of 100 3D printers, eventually rising to 1,000, as part of an integrated logistics solution close to the air hubs. Connected online, they can produce 3D-printed items using a wide range of technologies, from SLA to SLS and, more recently, Continuous Liquid Interface Production (CLIP), as well as more traditional techniques like CNC machining and injection molding. Customers can send designs directly to these printers, and the parts are then packaged and shipped via UPS's distribution network. With this complementary set of capabilities, a designer sending their data file to the facility by the 1:00 a.m. pickup time can have 100 copies of the item delivered anywhere in the US the next day,[33] and, once the Singapore facility is fully operational, within 24 hours to most major Asian cities. Such has been the success of Fast Radius's model that the firm was named "one of the top nine factories in the world" by the World Economic Forum in September 2018.[34]

However, there will be a limit to what 3D printing can do. Given the costs of materials, printing time, postprocessing, and the lead time to actually make items, there will always be a point at which injection molding and other mass-production technologies will be cheaper to use per unit (Figure 2.6), even if that cost point is dropping.

For now, we are several years, if not decades, from seeing the sorts of costs and speeds that will make 3D-printing plastic spoons en masse a better option than injection molding. While most of today's products are made in the hundreds of thousands, particularly those aimed at consumers, 3D printing only becomes a competitive option with a much smaller quantity of items.

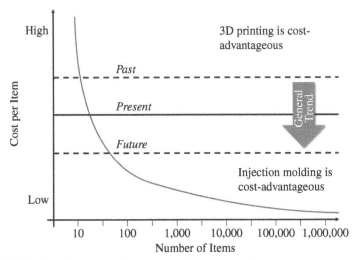

FIGURE 2.6   Development of cost per unit versus volume.

3D printing can still be useful in those mass-production scenarios, however: for instance, it reduces the time and cost to produce the molds for injection, and is thus preferable to using traditional techniques. The company Whale Pumps, for example, has found that 3D-printed injection molds are produced 97% faster than traditionally machined versions, something that several other firms have seen. Even with the higher price for 3D printing materials, companies are achieving cost savings of between 40 and 70%. The capabilities of 3D printing also allow such molds to include features that were previously unavailable or too expensive, such as efficient internal cooling channels, something that has led companies like Lego, which relies on injection molding, to use it.

For a long time, 3D printing had been less attractive for larger objects. Doubling the size of an item increases the number of points that need to be 3D printed by a factor of eight. However, the size of parts that can be 3D printed is increasing. While many people's first encounter with 3D printers will be via desktop models, or even photocopier-sized machines in printer hubs, some specialist firms have used 3D printing techniques to make large items, from cars to houses. Companies such as Winsun Decoration Design Engineering are aiming to 3D-print entire living quarters, and many organizations in the humanitarian aid sector hope to use this technology to build better emergency shelters following disasters. Winsun layers quick-drying cement and recycled raw materials through 3D printers to build small houses in under a day, at a cost of under US$ 5,000 each. Their process makes blocks offsite that are then assembled where needed, and the technique is scalable for larger structures. In Dubai, the company Convrgnt Value Engineering was contracted to develop a 3D-printed laboratory, and the city has pledged that 25% of its building projects will use this method by 2030. Dubai is also the location of the first office building projects to be 3D printed, built by 17 people in 18 days.[35]

▓ FINISH

As mentioned earlier, for much of its past, 3D printing was used to make prototypes and models, neither of which were expected to have the best finish. As it moves from being a tool for designers to a production manufacturing technique, more emphasis is being placed on the end product's finish. Extrusion-based items demonstrate this clearly, often displaying rough surfaces that need to be coated or machined smooth. In the right circumstances, 3D-printed items can be directly manufactured for immediate use, with little postprocessing—usually where a high level of finish is not needed. In June 2013, the Hong Kong branch of the toy store Toys "R" Us used 3D Systems Cube 3D printers to offer customers little ducks which were presented in small Dim Sum baskets with attached, personalized name plates. All three of those items were made on the 3D printers which had the necessary quality to make them sellable directly with no finishing steps needed. In most cases, though, and despite the apparent assurances from many 3D-printing enthusiasts, most products require some level of postmanufacturing processing and finishing before use. If supply chains are going to employ 3D printing, then those steps need to be factored into schedules and costs. That starts with identifying the object's end-use needs.

The condition in which a 3D-printed item comes out of the printer is a function of several factors, from the technology being employed, the materials being used, to the quality of the equipment itself—all of which directly correlate with the price of that equipment. For instance, if metal powders are exposed to humidity before the 3D-printing process begins, they may oxidize, which changes their characteristics, which in turn affects how they behave when submitted to lasers. Their condition also depends on the expertise of the installer and operator, the stability of the 3D printer, and the environment in which the item is made. Of course, 3D-printing technologies themselves have limitations: even the best laser-printing process is restricted in terms of its maximum precision, as the results from melting metal powders will not be of the highest levels without postprocessing.

More generally, postprocessing is a required stage of the entire 3D-printing process for end-use objects. For starters, in many 3D printing technologies, the items are fused to the base upon which they are made, or supports are added to designs for use during the 3D printing process; both will need to be removed mechanically (e.g. with wire cutting). Heating 3D-printed objects can release internal stresses that arise as a result of the fabrication process and that can warp objects if not removed. Consider the postprocessing needed to deal with complex designs. If using powder-based 3D printing such as laser (e.g. SLS, SLM, or Direct Metal Laser Sintering) or electron beams (e.g. EBM), surplus ("uncooked") powder needs to be removed from the object, a particularly difficult task if such material remains within the object itself. Even the specific type of technology used can hinder this: it is easier to remove uncooked powder from items made with lasers than it is if they are produced with electron beams. Many of these postprocessing techniques are very manual, contributing to an item's overall production cost and time to make. While removing uncooked powder can be achieved using vibration beds, many engineers prefer doing so by hand to reduce the risk of damage.

As with objects made using traditional techniques, items that are 3D printed also usually require some level of aesthetic finishing, such as CNC milling, treating (e.g. using isostatic pressing) and surface finishing (e.g. coating, polishing, and painting). This is particularly necessary when the finest tolerances are required, such as for dental implants and highly engineered aerospace and automotive parts. While much can be done with modern 3D printers, there is a drawback: the finer the tolerance needed, the thinner the layers needed to make the area or item, which increases the build time and production cost. Instead, it is preferable to use secondary machining to achieve those tolerances, something typically done using CNC machining or five-axis milling. Other finishing techniques have shown to improve the characteristics of 3D-printed parts. For example, the firm Graphite uses metal plating to increase the strength of SLA-printed ceramic parts, with thin layers of copper and nickel doubling the stiffness of items and increasing their strength by 40%.[36]

As mentioned in Chapter 1, postprocessing and finishing must not be underestimated—they can easily account for 75% of the effort to manufacture an item.

COST

The issue of the cost of 3D printing is not simple. On the one hand, there are capital and operational expenditure elements such as equipment, materials, and other 3D-printing-specific requirements. On the other, there are economies in cycle times, setup costs, and savings that 3D printing enables. Any consideration of 3D printing must look at both, taking a wider, holistic view.

3D printers that can be used for industrial goods have long been expensive, requiring capital investment of several thousands, if not hundreds of thousands, of dollars. While rising competition between hardware manufacturers has helped to drive down prices, models capable of producing high-quality objects, particularly in specialist metals and other exotic materials, continue to be costly. A Stratasys Fortus 900mc Production System, which makes objects in thermoplastics at the highest levels of accuracy and throughput, can cost over US$ 200,000, and an EOS M 290 SLS/SLM 3D printer over US$ 500,000, both at 2018 prices. As well as the printer itself, the cost is likewise a function of the materials that it uses: printers that produce objects in titanium have price tags in the order of hundreds of thousands of dollars, while desktop printers that use thermoplastics aimed at hobbyists cost a few hundred. Making the equipment itself cheaper is a tough challenge; for instance, SLS printers employ industrial lasers, sensitive thermal controls, and high-specification sensors, all of which need to operate at high temperatures for prolonged periods. That will never be cheap to accomplish, although low-cost models are coming on to the market at around US$ 5,000 but they sacrifice some capability to achieve that. Rather than investing in expensive equipment, there are alternative approaches to acquire 3D printing capabilities, and these will be explored in more detail later in Chapter 4.

The biggest contribution to operational expenditure of 3D printing, and one of the biggest factors holding back its wider adoption, is the cost of raw materials. A kilogram of plastic polymer material can cost anywhere from 2 to 104 times that of an equivalent plastic for injection molding or use in reductive manufacture, and metal powders destined for FDM or SLS can cost anywhere from 7 to 15 times the equivalent base material by weight.[37] This inflation stems from the complex processes needed to produce materials with the right characteristics for 3D printing. In situations where 3D printers result in production failures, this higher material cost can quickly become very expensive. However, it is a well-recognized issue in the 3D printing and materials sectors, and with improved materials research and manufacturing processes, and increased competition, the prognosis is for costs to fall.

Despite these apparently high costs, when the full end-to-end cost of 3D-printing an object is assessed, and the costs of design, set up, logistics and waste are also considered, 3D printing is fast becoming competitive with traditional manufacture, if not already cheaper in many cases, and this will accelerate its adoption. The Alliance for American Manufacturing has highlighted that, as prices drop, adoption will be fastest with small manufacturers, especially tool and die companies that can use the technology to ramp up production much faster and far more cheaply than with traditional tooling, with its high capital expenditure.[38]

Already, the savings accrued through the reduced number of manufacturing stages, the shorter cycle times from design to manufacturing, and the savings of lesser tooling requirements are considerable. Schneider Electric has reported that they have reduced the cost of producing injection molds by 90%, from €1,000 to €100, by switching from traditional to 3D printing manufacture. FDM reduces the cost of mold production by between 50 and 70%, with lead times similarly shrunk by 60 to 80%.[39] The company Divergent 3D, which makes cars using carbon fiber structures and DMLS of aluminum, has assessed the possibility of significantly reducing the tooling and manufacturing costs associated with making cars. When the reduced development time and capital investment are included, they find that vehicles can be made 20 to 50 times more cheaply than using traditional automotive production methods.[40] As a result, the company advocates the establishment and operation of 3D micro-factories, costing some US$ 4 million each, as opposed to the US$ 500 million to $1 billion price tag for a normal car plant. In September 2016, Divergent 3D announced that it had partnered with PSA Group, which owns Peugeot and Citroën, opening the possibility of a major shift in auto manufacture.

Another factor to consider when comparing the costs of 3D printing versus traditional techniques is complexity. As the complexity rises, as complexity rises, so does the cost to make an object, driven by the need for more specialized tools, sophisticated techniques, and stages in the manufacturing process. With 3D printing, changing complexity has few cost implications: making a highly complex item is as onerous as making a simple one, and the costs of doing so are relatively constant (Figure 2.7). The only differences come from greater energy consumption, more complex postprocessing and finishing, and higher materials usage stemming from the need for supports and perhaps higher wastage.

As 3D printing matures, so the cost per item has been falling, lowering the threshold where 3D printing is more economic for complex items.

Cost savings may also come from the wider disruption that 3D printing enables. For instance, reducing the weight of objects on an aircraft reduces

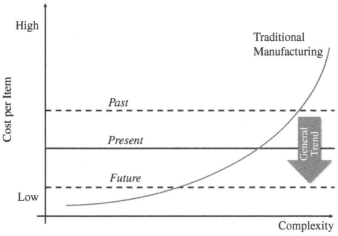

FIGURE 2.7   Comparison of 3D printing versus traditional manufacturing based on complexity.

fuel consumption, which reduces the costs per flight—this is significant when one considers that 30% of the cost of a flight comes from fuel. In the medical field, the costs of operations are being lowered by using 3D-printed models to assist surgeons in carrying out preoperative assessments in reconstructive and other forms of surgery. A 2005 study of 100 US hospitals found that the costs of operating rooms averaged US$ 62 per minute (typically, the cost ranges from US$ 22 to $ 133 per minute).[41] Traditionally, surgeons use film-based radiographs to understand the situation inside a patient and plan the surgery. Using 3D-printed models to acquaint surgeons and help train them in their forthcoming operations has been shown to reduce the lengths of those procedures by 25%,[42] thus saving some US$ 2,700 per operation purely in operation fees. Moreover, the reduced operating time results in shorter hospital stays and thus lower risks of postoperative infection, further reducing end-to-end medical costs. Patients whose surgeons have employed such 3D-printed models also have less pain during their postoperative care and less scarring. When one considers that, in Cherkasskiy and associates' study, the printer used for the models cost US$ 2,200 and each model US$ 10, the return on investment is undeniably clear.

The Applicability of 3D Printing

3D printing began its life as a way to make prototypes more quickly than with traditional manufacture, and it is still used for those. Now that the technology is maturing and moving further down the product value chain, what else can 3D printing be useful for? Usually, the immediate response refers to end-use items, whether assembly components, entire single objects, or spare parts, all aimed at some user to employ or consume. This includes Sonova's hearing aid casings, GE's fuel nozzles, Ford's spare parts, and the tools that Made in Space's printer makes on the International Space Station. As highlighted in this chapter, 3D printing is unlikely to be more economical than injection molding and other mass-production techniques. However, one often overlooked group of items are those used in traditional manufacturing processes. While 3D printing will never be able to mass-produce plastic teaspoons more cheaply and quickly than injection molding, it will make the molds used to produce them. These manufacturing tools are most commonly made using CNC machining, which, although precise and reliable, is an expensive technique that involves long manufacturing lead times.

The characteristics of 3D printing, particularly the ability to make items with complex geometries, make it very suitable to several manufacturing processes:

- Molding, including all varieties, such as injection, blow, liquid silicone rubber, and room temperature vulcanization, as well as those processes that use soluble cores for hollow composite parts
- Casting, including sand, spin, and investment processes
- Forming, such as thermoforming and metal hydroforming.
- Machining, including producing the jogs, fixtures, modular fixtures, and so on that are necessary
- Robotics end-effectors, such as the grippers used in robotized manufacture

Leading and specialist makers of tools for these processes are already using 3D printing. The engineering company Schneider Electric has embraced this in its Openlab model shop facility in Grenoble, France. The company uses FDM and Polyjet 3D printers to produce injection molds at a cost per unit an order of magnitude cheaper than a traditional aluminum mold, and with a production lead time of one week rather than one or two months.[43] In another example, VW Autoeuropa has been 3D-printing liftgate badge gauges, which accurately and precisely place the name of models on the rear of a vehicle. The cost of making each gauge using a low-cost, plastic FDM 3D printer is €10 as opposed to the €400 that VW was paying to have them made using traditional machining. Moreover, the lead time for a new gauge is reduced from 35 to 4 days.[44]

LOCATION

One of the greatest supply chain advantages of 3D printing lies in the flexibility of where to place production. Traditionally, manufacturing has been located in buildings large enough to accommodate the machinery and all of its support, and siting is often a balance of access to raw materials, energy generation, transportation channels, proximity to suppliers or customers, local fiscal incentives, and many other factors. 3D printing brings far greater flexibility than has been previously possible. With its smaller physical footprint, it removes many of the constraints, placing manufacturing where it is best to do so. This can be at the manufacturer's location, of course, but it could also be distributed anywhere in the world, and even at the customer's own premises. It can be outsourced more easily, releasing the need for significant CAPEX in facilities. This has the significant advantage of shortening the logistics chain and accelerating delivery lead times, a theme that will be discussed in more detail later in this book.

3D printing can also be located in places unsuitable for traditional manufacturing plants, as shown by the earlier example of the International Space Station. This flexibility is of extreme interest to armed forces around the world, for instance, who are looking to put 3D printers on ships and submarines, and with units deployed in the field. Already the US Navy has installed 3D printers on the amphibious assault ship USS *Essex*, enabling it to make entire unmanned aerial vehicles (UAVs),[45] each with rapidly configurable designs that render them suitable to the specifics of the mission. If used on submarines, 3D printing allows onboard engineers and technicians to make needed tools and spare parts without incurring lengthy detours to port or resupply at sea, increasing their time on station. In the healthcare sector, pharmaceutical companies are looking to deploy 3D printers at hospitals, closer to patients, so that they can not only produce the necessary tablets that that patient needs, but to do so with the exact dosage that they need, moving away from the "one size fits all" approach that currently exists.[46]

The flexibility of locating 3D printing is coaxing innovative solutions in other sectors. The concept of distributed manufacture, where items are made where it is most beneficial to do so, is now achievable, and the twin advantages of proximity to customers reducing delivery lead times, and minimizing or avoiding trans-border costs are now more readily achievable.

Locating production more widely also increases supply chain responsiveness, enabling manufacture to adjust to changing demand faster and more cheaply than with traditional manufacture. Normally, if the design of an object being manufactured is altered, production lines need to be adjusted and retooled, which can mean stopping production to effect those changes. If those changes need to be implemented quickly or production cannot be halted, then the supply chain will need to invest in a second production line. By contrast, 3D printing has the ability to absorb changes to design at no extra cost to the machinery involved. If the design changes require a change in 3D printing technology, the time to bring the new production capability online is reduced, and this can be minimized by using third-party-owned 3D printers. If required volumes ramp up, then they can be made wherever there is capacity for those items, using internally owned and managed 3D printers or pushing production to outsourced providers, which mitigates the need for and size of further capital investment (of course, logistics must be able to respond to delivery targets).

Where shorter times to delivery are needed, manufacture could conceivably take place during transit time, say, by 3D printing in the delivery vehicle. In 2015, Amazon filed a patent for placing 3D printers on trucks that make parts and products en route to the customer, allowing for even faster delivery, with no need for customers to invest in 3D printing directly themselves, or for the trucks to park at or near to the customer's base while acting as a mobile manufacturing plant.

Other configurations that place 3D printing close to raw materials can benefit supply chains by reducing the times and complexity of inbound logistics, and slashing risk by spreading manufacturing across several sites, mitigating major disruptions due to natural disasters. On May 2, 2018, Ford were compelled to halt production of the F150 pickup truck, one of its biggest sellers, because of a fire at the facility of one of its suppliers, Meridian Lightweight Technologies, in Easton Rapids, Michigan. That supplier provided Ford with several critical components used to make the vehicles, including the magnesium radiator support structures that hold radiators on the trucks. More importantly, that supplier and that facility were the only ones available to Ford in the USA.[47] Had that item been 3D printed instead, production could have switched to an alternative location faster and more easily, minimizing its impact on production with little need to duplicate the production facilities (often a traditional manufacturing risk mitigation strategy).

While the location where 3D printing can take place is far more flexible, there are limitations—we can't just print anywhere. 3D printers are sensitive equipment, particularly when producing objects with high precision. That means that they need to be situated on stable foundations, free from violent movements and vibrations, which precludes many locations. For example, although the concept of placing 3D printers on ships and submarines seems attractive, Julie Christodoulou, director of the Material Science and Technology Division at the US Office of Naval research, explains:

> "Factors such as water vapor and ship movement, make additive manufacturing in such environments more difficult."[48]

While shock absorption and insulation has long been understood and employed with other sensitive equipment, this is an emerging area for 3D printing, and placing the printers on things that move will be constrained in the short term.

LABOR

As all manufacturing does, 3D printing requires specialist operators to set equipment to work and operate it. It needs highly trained designers, engineers, and technicians throughout its value chain, as explored in more detail in Chapter 8. When compared to traditional manufacturing, 3D printing involves far fewer people to achieve the same results. Traditional manufacture typically requires significant resources to build a manufacturing line and set it to work, as well as to manage and tend to that line during production, consuming further resources for postproduction processes and finishing. 3D printing is simpler: once delivered, 3D printers require a few technicians to configure and mobilize them, but once in operation, they can be pretty much left to itself, needing a smaller crew (far less than in traditional manufacture) to keep them operational and deal with issues during printing. For instance, Fast Radius's production facility requires only three employees, each working an eight-hour shift in the plant, which is mostly "lights out"; that is, operating in the dark, something that has been achieved quickly and simply with high levels of automation.

What this will do to the labor market is change its shape. 3D printing undoubtedly threatens lower-skilled, low wage manufacturing jobs, calling for operators with higher skills levels. For instance, obtaining the right quality of objects from 3D printing relies on ensuring that the 3D printers are installed, operated, and maintained correctly. That last requirement—the need for suitably qualified maintenance personnel—is frequently forgotten when supply chains evaluate the viability of 3D printing. Whether jobs that were previously sent overseas will be re-shored as well, brought back to companies' home countries, will depend on the operating model that supply chains employ.

This need for different amounts of labor can become a significant issue. As it changes ways of working and the numbers of people involved, there may be restrictions on the scale and speed of 3D printing adoption in those countries where workers' rights are more stringently protected. Certainly, some engineers and technicians will be retrained, but the lower labor needs of 3D printing inevitably means fewer people. This naturally benefits the supply chain by reducing personnel costs.

SUSTAINABILITY

With corporate social responsibility now a growing factor in the decisions companies make, understanding the "green" credentials of supply chains is increasingly an important element of their success. No longer can companies operate with impunity, with no comprehension of where their materials come from, not knowing the social and environmental consequences of their operations. For instance, the jewelry sector must mitigate the risk of blood diamonds (i.e. those sourced from conflict zones) entering its supply chain. The mobile phone sector and the confectionery industry have to provide assurance that their raw materials—rare earth metals and cocoa, respectively—don't employ slave or child labor. The home appliance sector has had to provide assurances of its energy efficiencies. Understanding the end-to-end green credentials is now a "need to have," not a "nice to have." So how "green" is 3D printing?

To answer that question, 3D printing needs to be viewed along several dimensions:

- Waste
- Recyclability
- Lengthening operational lives
- Energy

Waste

The reductive manufacturing techniques of filing, milling, or drilling all entail the removal of unneeded material from an initial volume, and therefore they result in a high degree of waste. While in some sectors where raw material prices are very high—gold, silver, and platinum, for instance—waste in the manufacturing process is carefully minimized, this can nonetheless still be considerable. With other materials, wastage levels can be very significant, with some estimates of up to 80% of the raw materials used being wasted in the manufacturing process. In contrast, because 3D printing predominantly uses only the material that it needs for the object that is being made, waste is typically under 3%. Of course, consideration of waste must include the needs for temporary supports during the manufacturing process, which may have to be disposable, as well as the risk of production failure during the 3D printing process resulting in rejected items. However, even taking such factors into account, overall levels of material waste are still hugely reduced. In another example, the sports equipment company Nike has reported that it has reduced waste in the manufacture of its high-tech FlyKnit sports shoes to a "thimbleful."[49]

As well as with raw materials, there are other sources of waste in the supply chain, such as the production of wrong or surplus items. Many traditionally made objects are fabricated in advance of the demand for them (e.g. such as with make-to-stock policies), and where that demand falls away or never materializes, products either have to be recycled or reused, or, if not possible, disposed of in landfills or incinerated. Despite significant investment in demand forecasting and market research to ensure that only what is needed is actually made, some 30% of everything manufactured is deemed to be waste within months of rolling off the production line.[50] By 3D printing only those items that are needed when they are needed—a true make-to-order approach—this surplus production is eliminated, rendering a much more sustainable supply chain. According to Christoph Schell, President Americas Region at HP, this was one of the drivers for motor manufacturer Audi's decision to implement 3D printing for its parts catalogue, where it has reduced the overproduction of certain parts.[51]

However, the advantageous characteristics of 3D printing can, in certain circumstances, result in more waste in the item's cradle-to-grave lifecycle. Consider the situation where an item assembled from many different parts is instead made in one go using 3D printing, as described earlier in this chapter. In that situation, the entire item becomes the lowest replaceable part rather than an element of it, so if part of that item requires replacement, the entire item needs to be replaced, thereby wasting the rest of it. Considering the volumes of raw materials and the amount of energy needed to make such items, more of these resources will be wasted in the newly redesigned item than if it were made of several smaller components. The challenge then becomes one of design, balancing the advantages of simplified assembly with future repair needs.

Recyclability

In parallel with materials minimization, the use of recycled materials is increasingly possible as more initiatives to develop those as feedstock emerge.

FIGURE 2.8 An umbilical clamp produced from recycled water bottles by ReFabDar.
Source: Photo by Alessandra Silva.

For instance, in 2015 the pilot program ReFabDar began a series of efforts to use some of the 400 tons of plastic waste generated daily in Dar es Salaam, making it suitable for 3D printing purposes. As well as improving green efforts, the venture aimed to open opportunities for entrepreneurs and manufacture in Tanzania. The plastic materials were used to 3D-print medical devices (Figure 2.8), spare parts, education items, consumer goods, and even jewelry.

This ability to recycle raw materials is one of the factors that has prompted many in the environmental sector to consider how 3D printing can contribute to realizing ambitions for a circular or a closed-loop economy. In this model, resources are kept within the production cycle for as long as possible, rather than being continuously fed from nonrenewable sources; ReFabDar follows this practice. Of course, far more progress is needed in areas such as materials science and waste management, but from a manufacturing perspective, the flexibilities and capabilities of 3D printing would point favorably at it being a component of such a system, and several research efforts are underway to investigate what is needed to convert that potential.

Longer Operational Lives

Another way in which 3D printing improves sustainability is in extending the life of items, postponing the need for their replacement due to a lack of spare parts. In sectors such as oil and gas, mining, defense, and aerospace, equipment may be in use for decades, and often it has to be replaced in its entirety because a component is no longer available, even if the remainder of the equipment is perfectly operational. By recreating the required parts on demand through reverse engineering and 3D printing, that equipment's lifecycle can be extended, reducing the need for complete replacement. Naturally, this cannot be done for complex parts involving multiple materials. Such limitations mean that decisions on whether to 3D print spares must take a holistic view of what is needed, as Jan Holstrom and Timothy Gutowski point out:

> "This practice is constrained by the large number of spare parts that cannot be produced using additive manufacturing (AM) without a significant effort in redesigning parts for AM and on-demand production."[52]

They go on to add that:

> "For most products in use, it is likely impossible to redesign all parts for on-demand production."

It is therefore incumbent on designers to look ahead and consider the option for such on-demand 3D printing of component parts to help in extending the life of the equipment that they are destined for. Without that, the benefits of extending equipment lifecycles may not be as great as they might first appear.

Energy

The energy consumption of 3D printing technologies is frequently highlighted as a significant financial and environmental benefit. Several studies have investigated this question over recent years and the general consensus is that, while 3D printing appears to be more energy efficient, it is too early to tell.* This is not helped by the different energy requirements of the various 3D printing techniques, particularly if considering the energy needs to manufacture the raw materials as well as to power the actual 3D printing operation. Certainly, earlier models of 3D printers using FDM and SLS techniques, which use lasers and electron beams, were significantly power hungry, more so than their traditional equivalents, but the evolutionary trend has been to reduce that energy consumption. When viewed from an end-to-end standpoint by considering the energy consumption to produce raw materials, the machinery itself,

---

* Examples of recent studies include those by Tao Peng of Zhejiang University, China, and Yiran Yang et al. of the University of Illinois, USA (notes 51 and 53, respectively).

and the manufacturing processes and distribution, the energy consumption of a 3D-printing option is apparently lower than a traditional one. There is less equipment needed with a 3D-printing approach than with a traditional machine, particularly when considering the simplification of supply and value chains that 3D printing enables. The US Department of Energy estimated that 3D printing can potentially reduce energy costs by 50% in the medium term, and cut material costs by 90%.[53] With traditional manufacturing, for example, customization was very energy intensive, whereas there is no additional energy footprint to 3D-print a single design of an object or a thousand variations of its design. As energy efficiency is a major factor of 3D printing's sustainability credentials, the focus should be on utilization, keeping the 3D printer actually printing more, thereby improving contribution of energy to a produced item's cost basis.

The abilities and applications of 3D printing lend themselves to other benefits in energy savings. For instance, 3D printing enables new designs of heat exchangers, one of the biggest sources of energy inefficiency in much equipment. The precision and flexibility of the technology allow for more efficient designs to be built, using less material than traditionally: some of the latest models are 20% lighter, 20% more effective, and manufactured in a fraction of the time compared to current designs.[54] The net effect of these savings translates into environmental and financial gains for the supply chain.

As will be discussed later in this book, 3D printing simplifies supply chains, reducing the size of manufacturing footprints and eliminating the need for multiple assembly plants—and the transportation and logistics between them. That simplification accrues considerable environmental and financial savings. Researchers Malte Gebler, Anton Uiterkamp, and Cindy Visser modeled the impact of 3D printing on carbon emissions, reporting their findings in their 2014 *Energy Policy Journal* article.[55] They concluded that, by 2025, the use of 3D printing can reduce supply chain costs by US$ 170 billion–$ 593 billion, with a saving in total primary energy supply needs of 2.54 to 9.30 EJ, and a combined lowering of CO_2 emissions of 130.5 to 525.5 Mt; that is the equivalent of taking half of all passenger vehicles in the USA off the roads immediately.

The response to the earlier question (how "green" is 3D printing) is that 3D printing is by far a greener alternative to traditional manufacturing processes when viewed from a systems perspective, considering all the contributions to the supply chain. More effort is needed to evaluate the scale of those impacts. For instance, the extraction, processing, and preparation of the many raw materials all have their own individual environmental footprints; each comes

A Greener Option—The Airbus Example

As part of its strategic goals, the aerospace company Airbus declared its intention to drive environmental improvements in its business, both within its own operations and in the products that it manufactures. The company's Innovations Group established a collaboration with the 3D printing technology firm EOS in 2014 with the aim of providing a tangible environmental case for using 3D printing techniques, specifically EOS's DMLS methodology.[56] The base case for comparison was rapid investment casting, a traditional if modern manufacturing technique.

The study first identified a small set of parts around which to analyze the data, and the first part chosen was the nacelle hinge bracket for an Airbus A320. To be credible, the analysis considered the full end-to-end production cycle for that part, from production of the raw material powders through the manufacturing to the end user. In this case, they analyzed the installed and operating hinge bracket, accounting for the composition, parts, processes and subsequent operation involved in its lifecycle. The results were produced using Airbus's tried and tested assessment approach, the Streamlined Lifecycle Assessment (SLCA) and the internationally recognized ISO 14040 series data. In the traditional manufacturing case, the part was cast in steel. With the DMLS approach, the part was made in titanium, with an optimized topology.

The study produced three important findings:

- Over that complete lifecycle, hinges made using DMLS with optimized topology resulted in a reduction of 40% in CO_2 emissions as a direct consequence of the lower weight in the finished part. When this was aggregated across all brackets on a single aircraft, it resulted in a weight reduction of 10 kg.
- There was a net saving in energy consumption by using the DMLS technique over casting. While the former is energy-hungry during melting and chilling, there was a significant reduction in the duration of that energy-intensive manufacturing. Casting, on the other hand, requires a hot furnace to burn a model that could be made in epoxy resin using SLA.
- The DMLS process required 25% less raw material than casting, with a reduction in postproduction machining.

For Airbus, the environmental credentials of 3D printing further build the case for its use in manufacturing parts for its aircraft. Indeed, such is the demand for manufacturing aerospace parts using 3D printing that the

(continued)

(continued)

company built a dedicated facility at the Ludwig Bolköw campus in Munich, Germany.

The pursuit of lighter parts is also driving car manufacturers to develop 3D-printed parts for future electric vehicles (EVs). The range of a vehicle, the time to charge, and several other performance factors that define the competitiveness of EVs are all functions of vehicle weight. With ever more governments overtly or indirectly pushing the automotive sector to shift to EVs, several of the biggest makers are now looking to use the capabilities to 3D printing to optimize topologies and lighten the load. GM, Ford, and Volkswagen have all publicly announced design programs to produce 3D-printed parts in their future EVs, opening the possibility of cheaper, faster, and more customizable replacement parts in those cars.

from a difference source, employing different types and sizes of labor force; each involves a different energy-consumption profile; and so on. These lifecycle evaluations are complex and dynamic, and are only now beginning to be carried out, both in the academic and commercial worlds.

THE CASE FOR 3D PRINTING

Aggregating the various aspects discussed in this chapter, the case for 3D printing is summarized in Table 2.1. For supply chains, these translate into increased customer satisfaction, greater flexibility, improved resilience, and better economics. All of these mean that it is making inroads into mainstream manufacture, away from just prototyping. While there continue to be constraints, these are being tackled and slowly resolved by academic and industrial R&D teams; in the meantime, users of 3D printing are working around them.

If we are to understand how the technology can be used to supercharge supply chains, it is also prudent to understand where that development is going, and to look at potential parallels with other technology developments, revising the lessons learned from those.

TABLE 2.1   The case for 3D printing.

Aspect	Benefit	Constraint
Speed	▪ End-to-end speed of 3D printing process is faster than traditional manufacturing	▪ Actual speed of 3D printing is slow
Materials	▪ Ever-increasing range of materials that can be 3D printed	▪ Fewer materials and material types compared to traditional manufacturing
Design	▪ Designs can be made with greater complexity than with traditional manufacturing ▪ Objects can be 3D printed with fewer parts ▪ Mass customization is viable	▪ Design process is more complicated than for traditional manufacturing
Design tools	▪ Design tools are starting to be integrated ▪ More capable file formats are now available	▪ Design tools are complicated and require training ▪ More capable file formats are not widespread
Volumes	▪ Lots of one are viable ▪ Volumes of production increasing	▪ Mass production is not possible
Finish	▪ 3D printers are producing higher-quality products	▪ Postprocessing and finishing needed, particularly for low tolerances
Cost	▪ End-to-end costs of production reduced by 3D printing ▪ Costs of equipment and materials are dropping	▪ 3D printers are expensive ▪ Materials for 3D printing are more expensive than traditional equivalents
Location	▪ 3D printing has more flexibility in where it can be established	▪ 3D printers require stable platforms
Labor	▪ 3D printing requires less labor	▪ 3D printing requires higher skill level
Sustainability	▪ 3D printing incurs less waste ▪ 3D printing can use recycled materials ▪ 3D printing extends equipment lifecycles ▪ 3D printing reduces energy footprint	▪ More research is needed to quantify sustainability credentials, especially energy consumption

Notes

1. Richard Hague, "3D Printing—Don't Believe All the Hype," *The Engineer*, April 5, 2017, https://www.theengineer.co.uk/3d-printing-dont-believe-all-the-hype/.
2. Carrie Wyman, "Fender Rocks Amp Design, Slashes Time to Market by More Than 50% with Stratasys 3D Printed Prototypes," Stratasys, July 1, 2015, http://blog.stratasys.com/2015/07/01/fender-amps-3d-printing.
3. Nyshka Chandran, "3D Printing Just Made Space Travel Cheaper," CNBC, June 7, 2015, http://www.cnbc.com/2015/06/07/3d-printing-made-space-travel-cheaper.html.
4. Ben Rossi, "Supply Chains Are Looking Better in 3D," Raconteur, February 23, 2018, https://www.raconteur.net/manufacturing/supply-chains-looking-better-3d.
5. T. E. Halterman, "Canadian Firm Bringing 3D Printed Evidence to the Courtroom," 3dprint.com, June 5, 2015, https://3dprint.com/71040/3d-printed-evidence-2.
6. ABB, "Plastic Spoons, deSter, Belgium. Case Study: Injection Moulding," 2006, https://library.e.abb.com/public/90e8bb3319adcc76c12577c70 02f5c7d/deSter_casestudy_en_print.pdf.
7. Zacks Equity Research, "Alcoa (AA) Inaugurates 3D Printing Metal Powder Facility," Yahoo!, July 6, 2016, http://finance.yahoo.com/news/alcoa-aa-inaugurates-3d-printing-223410393.html.
8. Organovo, "L'Oréal USA Announces Research Partnership with Organovo to Develop 3-D Bioprinted Skin Tissue," May 5, 2015, http://ir.organovo .com/phoenix.zhtml?c=254194&p=irol-newsArticle&ID=2129344.
9. Wake Forest Institute for Regenerative Medicine, "Using Ink-Jet Technology to Print Organs and Tissue," Wake Forest School of Medicine, 2016, http://www.wakehealth.edu/Research/WFIRM/Our-Story/Inside-the-Lab/Bioprinting.htm.
10. Maddy Savage, "The Firm That Can 3d Print Human Body Parts," BBC, November 15, 2017, http://www.bbc.co.uk/news/business-41859942.
11. Katia Moskvitch, "3D-Bioprinted Ears Could Be 'Five Years Away.'" Institution of Mechanical Engineers, November 9, 2017, http://www.imeche .org/news/news-article/3d-bioprinted-ears-could-be-'five-years-away'.
12. Jasper Eddison, "3D Bio-Printing: A Medical Revolution?" *Berkeley Squares*, November 13, 2017, https://www.berkeleysquares.co.uk/2017/11/3d-bio-printing-a-medical-revolution.

13. J. Gooch, "3D Printing Steals the Show at Euromold 2015," Digital Engineering, October 2, 2015, http://www.rapidreadytech.com/2015/10/3d-printing-steals-the-show-at-euromold-2015.

14. NASA, "NASA Tests Limits of 3-D Printing with Powerful Rocket Engine Check," August 27, 2013, https://www.nasa.gov/exploration/systems/sls/3d-printed-rocket-injector.html.

15. "Finding Flaws in 3D-Printed Titanium," Medical Design Briefs, July 1, 2016, http://www.medicaldesignbriefs.com/component/content/article/mdb/tech-briefs/24997.

16. Ann R. Thryft, "3D Printing Adds Value All Along the Value Chain," *Design News*, August 24, 2016, https://www.designnews.com/content/3d-printing-adds-value-all-along-value-chain/186825924845367.

17. "For Aircraft Engine Bracket, 3DP Completes the Optimization," *American Machinist*, March 2, 2016, http://americanmachinist.com/cadcam-software/aircraft-engine-bracket-3dp-completes-optimization.

18. Maté Petrány, "Bugatti Can Now 3D-Print This Gorgeous Brake Caliper from Titanium," *Road & Track*, January 22, 2018, https://www.roadandtrack.com/new-cars/car-technology/a15840339/bugatti-can-now-3d-print-this-gorgeous-brake-caliper-from-titanium.

19. Oliver Wainwright, "The First 3D-Printed Pill Opens Up a World of Downloadable Medicine," *The Guardian*, August 5, 2015, https://www.theguardian.com/artanddesign/architecture-design-blog/2015/aug/05/the-first-3d-printed-pill-opens-up-a-world-of-downloadable-medicine.

20. Lee-Bath Nelson, "Industrial Strength—Revolutionary 3D Printed Metal Designs GE-Style," LEO Lane, October 7, 2015, https://www.leolane.com/blog/industrial-strength-revolutionary-3d-printed-metal-designs-ge-style/.

21. Tomas Kellner, "An Epiphany Of Disruption: GE Additive Chief Explains How 3D Printing Will Upend Manufacturing," GE Reports, November 13, 2017, https://www.ge.com/reports/epiphany-disruption-ge-additive-chief-explains-3d-printing-will-upend-manufacturing/.

22. Ibid.

23. Benedict, "GE's $200M 'Brilliant' 3D Printing Factory in Pune, India to Develop Critical End-Use Parts," 3ders.org, April 7, 2016, http://www.3ders.org/articles/20160407-ge-200-million-multi-modal-3d-printing-factory-pune-india-develop-critical-end-use-parts.html.

24. Tomas Kellner, "An Epiphany of Disruption: GE Additive Chief Explains How 3D Printing Will Upend Manufacturing," GE Reports, November 13, 2017, https://www.ge.com/reports/epiphany-disruption-ge-additive-chief-explains-3d-printing-will-upend-manufacturing/.

25. D Natives, "Top 10 3D Printing in Aeronautics," March 1, 2018, https://www.3dnatives.com/en/3d-printing-aeronautics-010320184.
26. Stuart Nathan, "Rolls-Royce Breaks Additive Record with Printed Trent-XWB Bearing," The Engineer, June 16, 2015, https://www.theengineer.co.uk/issues/june-2015-digi-issue/rolls-royce-breaks-additive-record-with-printed-trent-xwb-bearing/.
27. PwC and The Manufacturing Institute, 2014, *3D Printing and the New Shape of Industrial Manufacturing,* June 2014, http://www.themanufactur inginstitute.org/~/media/2D80B8EDCCB648BCB4B53BBAB26BED4B/3D_Printing.pdf.
28. Scott J. Grunewald, "Stratasys 3D Printers Helping to Launch Rockets into Space," 3dprint.com, April 20, 2015, http://3dprint.com/59428.
29. Dibya Chakravorty, "What Is a 3D Printer File Format?" all3dp.com, June 16, 2018, https://all3dp.com/3d-printing-file-formats.
30. Judith Vanderkay, "3MF Consortium Launches to Advance 3D Printing Technology," 3MF Consortium, April 30, 2015, http://3mf.io/3mf-consortium-launches-to-advance-3d-printing-technology/.
31. Katharine Sanderson, "3D Printing: The Future of Manufacturing Medicine?" *The Pharmaceutical Journal,* 294, no. 7865 (June 2, 2015), https://www.pharmaceutical-journal.com/news-and-analysis/features/3d-printing-the-future-of-manufacturing-medicine/20068625.article?firstPass=false.
32. PwC (2016). "3D Printing Comes of Age in US Industrial Manufacturing," accessed June 30, 2016, http://www.pwc.com/us/ en/industrial-products/3d-printing-comes-of-age.html.
33. "3D Printing to Disrupt the Future of Supply Chain," STAT Trade Times, accessed December 12, 2016, http://www.stattimes.com/blog/3d-printing-to-disrupting-the-future-of-supply-chain.
34. Cann, Oliver, Europe, Asia Lead the Way to the Factories of the Future, World Economic Forum, September 2018, http://www.weforum.org/press/2018/09/europe-asia-lead-the-way-to-the-factories-of-the-future.
35. Alicia Miller, "3D Printed Buildings Are Coming to Dubai," 3dprintingin-dustry.com, July 14, 2016, http://3dprintingindustry.com/news/3d-printed-buildings-coming-dubai-87203/.
36. "Metal Plating Benefits," Graphite Additive Manufacturing, March 22, 2017, http://www.graphite-am.co.uk/2017/03/metal-plating-benefits/#more-1292.
37. John J. Coyle and Kusumal Ruamsook, "T = MIC²: Game-Changing Trends and Supply Chain's 'New Normal,'" *CSCMP's Supply Chain*

Quarterly, September 24, 2018, http://www.supplychainquarterly.com/
topics/Strategy/20141230-t-mic2-game-changing-trends-and-supply-
chains-new-normal/.

38. PwC and The Manufacturing Institute, *3D Printing and the New Shape of Industrial Manufacturing*.

39. SMG3D, "Metal Hydroforming with a 3D Printer," n.d., https://www
.smg3d.co.uk/3d_printer_applications/metal_hydroforming_with_a_
3d_printer.

40. Derek Markham, "Divergent 3D Seeks to Radically 'Dematerialize' Auto Manufacturing with 3D-Printing & Microfactories," Treehugger, January 19, 2016, http://www.treehugger.com/cars/divergent-3d-seeks-radically-dematerialize-auto-manufacturing-aluminum-3d-printing .html.

41. Ronald D. Shippert, "A Study of Time-Dependent Operating Room Fees and How to Save $100,000 by Using Time-Saving Products," *American Journal of Cosmetic Surgery* 22, no. 1 (2005): 25–34.

42. L. Cherkasskiy, et al., "Patient-Specific 3D Models Aid Planning for Triplane Proximal Femoral Osteotomy in Slipped Capital Femoral Epiphysis," *Journal of Children's Orthopedics* 11 (2017): 147–53.

43. Daniel O'Conner, "The Factory of the Future Is Here and Now," *TCT Magazine*, November 23, 2016, https://www.tctmagazine.com/3d-printing-news/the-factory-of-the-future-is-here-and-now.

44. Daniel O'Conner, "Can You Jig It? 3D Printing inside Volkswagen Autoeuropa," *TCT Magazine*, February 22, 2018, https://www.tctmagazine.com/can-you-jig-it-volkswagen-ultimaker-3d-printing.

45. Martyn Williams, "The US Navy Is 3D-Printing Custom Drones on Its Ships, *PCWorld*, July 29. 2015, https://www.pcworld.com/article/2954732/the-us-navy-is-3dprinting-custom-drones-on-its-ships.html.

46. Wainwright, "The First 3D-Printed Pill."

47. Natasha Bach, "Why Ford Has Stopped Production of America's Best-Selling Pickup," *Fortune*, May 10, 2018, http://fortune.com/2018/05/10/ford-f150-production-halt-fire/.

48. Jon Harper, "Military 3D Printing Projects Face Challenges," *National Defense*, November 1, 2015, http://www.nationaldefensemagazine.org/.

49. Tess, "You'll Soon Be Able to 3D Print Nike Shoes at Your Home, Says Nike COO," 3ders.org, October 7, 2015, http://www.3ders.org/articles/20151007-youll-soon-be-able-to-3d-print-nike-shoes-at-your-home-says-nike-coo.html.

50. Filemon Schoffer, "Is 3D Printing the Next Industrial Revolution?" TechCrunch, February 26, 2016, https://techcrunch.com/2016/02/26/is-3d-printing-the-next-industrial-revolution/.

51. Christoph Schell, "3D Printing Will Reinvent the 'Rule of the Road' in Automotive Manufacturing," LinkedIn, November 29, 2016, https://www.linkedin.com/pulse/3d-printing-reinvent-rules-road-automotive-christoph-schell.

52. Jan Holstrom and Timothy Gutowski, "Additive Manufacturing in Operations and Supply Chain Management: No Sustainability Benefit or Virtuous Knock-On Opportunities?" *Journal of Industrial Ecology* 21, no. S1 (2017): S21.

53. Tao Peng, "Analysis of Energy Utilization in 3D Printing Processes," *Procedia CIRP* 40 (2016): 62–67.

54. Yiran Yang, Lin Li, Yayue Pan, and Zeyi Sun, "Energy Consumption Modelling of Stereolithography-Based Additive Manufacturing toward Environmental Sustainability," *Journal of Industrial Ecology* 21, no. S1 (2017), https://onlinelibrary.wiley.com/doi/full/10.1111/jiec.12589.

55. Malte Gebler, Anton J. M. Schoot Uiterkamp, and Cindy Visser, "A Global Sustainability Perspective on 3D Printing Technologies," *Energy Policy* 74: 158–67.

56. EOS, "Life Cycle Cooperation between EADS IW and EOS," October 2013, https://cdn0.scrvt.com/eos/public/3e6e0ce0e863a54d/61a83feac49c2e6ab2c12a48a2457f54/eos_study_en.pdf.

CHAPTER THREE

Where Is 3D Printing Heading?

D URING ITS FIRST two decades as a tool for rapid prototyping, the types and durability of materials, quality, cost, and speeds of 3D printing meant that it wasn't suitable for any other purpose. Engineers and designers came to embrace its ability to make prototypes for confirming form, fit, and function, or simply to have something to show people. Enthusiasts were more ambitious, and they set about thinking of what else could be done. Since then, the original set of materials has expanded, the range of technologies and their characteristics and capabilities have widened, and the competence of designers has deepened. The result is that 3D printing has stepped outside of its rapid prototyping sphere into mainstream manufacturing.

That 3D printing has already changed many industries is clear. It is an established production technique in sectors from transport and jewelery to dentistry and space, and more sectors are adopting it. To better realize its disruptive effect, significant improvement is needed in the three main technical areas of hardware, software, and materials. As those continue to mature, all engineering, manufacturing, and technology industries will be affected over the coming years if they haven't already been so. Given the billions of dollars in investment in these three areas, progress is practically assured sooner or later. Trying to divine how 3D printing will develop, and where and what its impact will be, requires looking at its past and current situation, and then comparing those to how other technologies have evolved.

The biggest change to affect the 3D printing market has been the shift of purpose. Increasingly, they are used for making jigs, fixtures, and injection molds for use with traditional manufacture, as well as considerable numbers of end-use items. Automotive manufacturer Volkswagen uses 3D printing to produce parts for the equipment that makes its vehicles, something that has saved the company over US$ 160,000, with over 1,000 such parts made in the first year of that project.[1] As this change of focus has taken hold, it has required a step change in the capabilities that have held 3D printing back for so long, from the ease of design to the precision and finish of parts made. This is significant as a business moves from start-up to full production: the capabilities and performance levels that got it going and were good enough in the early days are simply not enough to sustain it as it ramps up.

This chapter will examine the near-future developments in 3D printing and the areas where it will enter, expand and mature, leveraging lessons from the recent history of other similarly disruptive technologies. It will look at the wider developments that are shaping supply chains and how 3D printing will fit into those. This will serve as a guide to supply chain decision makers on what to expect in the future.

 ## INTEGRATION AND THE RISE OF THE ROBOTS

3D printing is a manufacturing technique with a series of stages, both virtual and physical. The virtual includes all the steps in the design phases, from initial drafting, through stress modeling, simulation, and so on. One of the often-heard complaints among 3D print designers is that they need to move data across several software packages, each of which is used for one or a small handful of those steps, and each package has its own look and feel. As well as slowing down the design stages, this exposes the data to error and breaches. The actual fabrication process involves a series of physical steps, from loading the 3D printers, preparing the printer beds, and moving the fabricated parts to the next stage and then onto a raft of postproduction steps. The hazards of 3D printing—explored later in this book—mean that the intervention and use of human operators further slows down the end-to-end cycle times, as they wait for chambers to be cleared of powders, high currents to be powered down, and the like. Each step takes time to complete, slowing down the process. 3D printer makers are increasingly addressing this, moving beyond simply developing the actual printers to providing wider integrated solutions, often in a modular way. With that approach, manufacturing supply chains will be scalable according to their needs, with a tangible gain in performance.

As mentioned in Chapter 2, design software providers are starting to unify the computer-aided design (CAD), computer-aided manufacture (CAM), computer-aided engineering (CAE), and other software platforms into single packages and common environments. Makers of 3D printers such as 3D Systems, EOS, and Additive Industries, are now integrating their offerings and incorporating automation in their newer releases. Additive Industries' MetalFAB1, for example, is a single system that integrates material loading, powder bed fusion 3D printing, and postproduction using a high level of automation. Once the design data is sent to the system, it fabricates them and moves them through the production stages in a seamless manner. This has the benefit of accelerating production as well as reducing risks to human operators; the production and postprocessing chain is shortened from two hours to six minutes, which also cuts the cost and increases the sustainability of manufacturing. Other 3D printer providers are working closely with robotics developers to assist with the loading of materials canisters, items that often weigh more than the recommended and legal limits for human workers to lift. Indeed, when combined, 3D printing and robotics offer supply chains a powerful combination: automated production that can operate with minimal human needs, reducing costs and disruptions, and therefore increasing productivity. This has opened up new capabilities during the production process; for instance, 3D Systems' Figure 4 Production series similarly coalesces the different stages of the 3D print process into a single unit and employs robotics to move materials and parts through it. Additionally, the firm has added a robotic arm to hold up a fabricated part to a laser scanner, which then checks that the finished item matches the geometry of the intended design within the acceptable tolerances.[2]

Those combined capabilities are being used by commercial companies in their operational models. Voodoo Manufacturing, for example, headquartered in Brooklyn, New York, built a 3D printing factory with over 200 machines that have already logged over a million printer hours, producing over 400,000 plastic parts; its biggest single order to date consisted of 22,000 items. Voodoo installed racks of fusion deposition modeling (FDM) 3D printers in their unit, offering customers two maximum sizes of printing envelopes, and producing items in volumes and at costs that challenge injection molding. To achieve this, Voodoo uses visually assisted, low-cost UR10 robotic arms placed in front of the racks that automate some of simpler but typically labor-intensive tasks, such as placing the print beds in the machines, removing parts once they're made—still on their beds—and placing them on a conveyor for postprocessing. This enables production to run continuously, with lights out, increasing productivity and reducing the costs per item.

▓ A DIGITAL LAUNCH PAD

In 1962, Sam Walton opened a store in Rogers, Arkansas, a small town in what was a quiet state. His assistant in this venture, Bob Bogle, who had been given the task of looking after signage for the store, suggested the name that quickly became known across the USA and beyond: Wal-Mart.[3] The company went public in 1970 and grew year-on-year, and by 1975 it had a network of 125 stores across the country, with a turnover of US$ 340 million, all based on the promise of low prices and great service. That year saw the company do something unexpected: it purchased an IBM 370/135 computer system to control its inventory across its warehouses and distribution centers. It also used the system to prepare income statements for each of its stores, previously a particularly laborious manual task. To help feed information into that system, it also installed electronic point of sale (EPOS) devices—cash registers—at over 100 of those stores, giving it visibility of what was selling and where it was being sold. This enhanced data allowed it to ramp up its sales, and within four years it was a billion-dollar firm, something it achieved in record time.

The success that Walmart (it removed the hyphen from its name in 2017 as it moved further into the online domain) had at using computing power and data analytics to transform its operations and, from those, its business performance, was instrumental in bringing IT into businesses in all industry sectors. Developments in technologies, from the advent of the Internet, data processing tools, and the cultural acceptance of digital innovations in operations slowly extended and deepened those changes. The early part of the twenty-first century then saw a shift in the pace of those changes. Today, organizations are at the center of a perfect storm of technologies and societal changes that is radically transforming their ways of working, driven by the ever-present needs for responsiveness, adaptability, and speed in their supply chain operations. At the heart of that transformation is the use of data to produce information and to make correct right decisions. This, then, is digital transformation—the change to operations, processes, and systems, to how people interact with them, and to culture through the use of data. The impact of this transformation has been emerging in different areas of the supply chain for several years:

- Supply chain planners who use data analytics to optimize where to place the elements of their operations or to develop algorithms that predict demand from a range of different signals
- Buyers who employ digital tools for e-procurement, running cloud-based reverse auctions to accelerate purchasing-cycle times while reducing prices
- Manufacturers who began to bring CAD designs into traditional production, such as with computer numerical control (CNC) machining and five-axes milling
- Logisticians who employ electronic picking assistants in warehouses, and GPS-enabled tracking devices to understand how their inventories are moving

All these developments are aimed at helping businesses better understand their customer needs and meet those requirements, to optimize supply chains at the best cost, and to bring control to those supply chains, enhancing the businesses' bottom lines.

To gather that data, the physical elements of supply chains, from machinery to vehicles, are increasingly fitted with sensors connected via computer networks. This is the so-called Internet of Things, which sees the prospect of billions of items, from cars to refrigerators, being connected online. Placing remote sensors in physical systems to collect information about their operation is not new, having been the norm for decades in most industrial sectors. Neither is the use of data analytics to provide insight, as seen in the Walmart example. What is different is the scale of the data that is produced and the speed with which that happens. By 2015, more data had been created in the preceding two years than in the whole history of humanity.[4] By 2017, 90% of the world's data—2.5 quintillion (2.5×10^{18}) bytes—had been created in the previous two years.[5] This is "Big Data," and it is changing how companies work and how they make decisions in all areas, particularly in the supply chain.

This is the latest step change in emerging technologies (Table 3.1) and how they're used, and it is leading to a revolution in industry. Over the last 200 years, manufacturing has undergone three such step changes in capability, each driven by technology (Figure 3.1):

1. In the early 1800s, the use of mechanization, powered by steam and water, enabled factories to transform the highly manual processes that they had

TABLE 3.1 Definitions of emerging technologies.

Technology	Definition
Digital transformation	The changes effected to human activity through the use of digital technologies
Internet of Things	The widespread use of sensors to gather data from physical systems via computer network
Big Data	Extremely large datasets that are analyzed to produce information on patterns, trends, and associations
Industry 4.0	The use of cyber-physical systems, the Internet of Things, cloud and cognitive computing in a production environment

used for nearly 1,000 years, significantly accelerating production times. This was particularly evident in the cotton mills of northern England and in agriculture globally. This was the so-called "Industry 1.0."

2. The beginning of the twentieth century brought mass manufacturing to factories (Industry 2.0), with pioneers like Henry Ford and William Klann improving the processes to increase volumes of manufacture.

3. The latter half of that century then saw the advent of computers and, more recently, automation in factories, realizing "Industry 3.0."

Now we are on the cusp of the next stage: Industry 4.0. This concept arose from the work of Siegfried Dais of Robert Bosch GmbH and Henning Kagermann of Acatech, who were initially commissioned by the German government.[6] Industry 4.0 brings all the previous advancements together with a systems engineering approach, producing cyber-physical systems—systems that seamlessly integrate digital and physical technologies. At first glance, this may not seem new: for decades, companies have been using sensors to produce data that is then analyzed to inform decisions and improve performance.

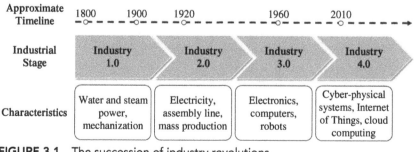

FIGURE 3.1 The succession of industry revolutions.

However, the depth, breadth, and scale of technologies throughout all areas of operations means that Industry 4.0 goes far beyond what was possible before, and those operations that employ it well become better utilized, characterized by highly flexible production and strong customization. Those aspects require manufacturing approaches that are equally highly flexible and allow for mass customization, beyond what has been achievable with traditional techniques; those are precisely the capabilities of 3D printing. With its lean production, which can be outsourced and automated, 3D printing is now seen as a key enabler of the new paradigm. To understand what this means, lessons can be learned from other recent technology developments, such as the advent of personal computers (PCs).

Although computers had been around since the late 1940s, it wasn't until the arrival of Steve Jobs, Steve Wozniak, and Bill Gates in the 1970s that they became PCs, entering businesses and homes in far greater numbers that they had till then. Despite their versatility and, at that time, advanced technology, PCs were seen initially as something for the hobbyist, for geeks and nerds, something to program to do "cool" things, from playing *Pong* and *Space Invaders* to calculating logarithmic tables quickly. Magazines dedicated to providing those hobbyists with programs, hints, and tips appeared, and a subculture soon emerged. That trend continued during the 1980s, with PCs slowly making headway in businesses, as well as within that geek community. However, as companies began to use them, they found that they were limited, that the realities of their capabilities did not match their promise. The machines were difficult to use, requiring training to code and run. It was not possible to exchange data between them without expensive infrastructure investment to lay dedicated cables. Then in 1989, something extraordinary happened: sales of PCs actually fell.[7] This year-on-year downward trend continued until a series of developments in the early 1990s. First, Microsoft released Windows 3 in 1990 and 3.1 in 1992, building on the more user-friendly but less commercially successful operating systems that rival company Apple offered, which made using computers easier. Around that time, inkjet and laser printers became more portable and affordable, allowing them to be easily installed in offices and homes. Alongside this, December 20, 1990 witnessed something that would go on to change so much: the debut of the first website on the newly created World Wide Web (WWW), which itself was opened to the public by its creators at CERN in Switzerland on April 30, 1993. The WWW allowed for communication between PCs on telephone wires, opening up the ability to network systems. The PC sector soon recovered its past growth trend, catalyzed by the coming together of all four of these developments.

3D printing looks like it is following a similar trajectory. There is now a tangible need for 3D printing, and the technical elements are coming together to fully realize its potential. The march toward digital transformation and the expansion in the use of Industry 4.0 solutions will catalyze the adoption of 3D printing in supply chains, in the same way that the advent of the easier-to-use software platforms, the Internet, and the availability of printers changed how office-based work could be delivered, propelling the adoption of PCs in all areas of business. It will open the digitization of sectors that previously have been held back from being transformed.

 ## SHIFTING DYNAMICS OF THE 3D PRINTING SECTOR

The hype of 3D printing has died down and there is far more pragmatism about what it is and what it can do. So, with 3D printing already a present element in many manufacturing process, what will catalyze its further adoption, helping it to achieve the potential that so many see for it? In the short term, the biggest innovations that will really ramp up the use of 3D printing lay in the area of materials: making items with several materials and in many colors in one go. The inclusion of conductive materials, such as copper, or more exotic materials, like Voxel8's conductive inkjets, within the structure of a 3D-printed object, thereby enabling transmission of electric currents or the embedding of electric circuits, will explode the number of uses for 3D printing. This is likely to happen soon, with the successes of 3D-printed carbon nanotubes already showing promise.[8]

In parallel, improving the economics of 3D printing will also be a critical factor. This was key to the adoption of other disruptive technologies, such as the PC and mobile phones, which took off when their price tags shrank significantly while their capabilities increased, offering what was perceived as good value for the money. This reduction in cost is likely to stem from increased competition between makers of 3D printers and the arrival of new entrants, as well as from cheaper materials, resultant from the expanded research and development in that space. To scale that, there were 49 manufacturers of 3D printing machines in 2014, which doubled to 97 in 2016.[9] Since the main patents lapsed, the costs of machines have reduced by a factor of 10, and their precision and accuracy have increased. As costs of 3D printing further drop, manufacturing will increasingly become a commodity, enabling small companies as well as large corporations to make objects quickly with low investments. Likewise, those firms that acquire the right makers of 3D printers, particularly for niche purposes, and turn them to their advantage will have a tangible competitive edge. This is driving mergers and acquisitions in the 3D printing space: just in 2016, GE acquired Arcam and SLM Solutions for US$ 14 billion, Siemens bought Material Solutions Ltd., and HP purchased David Vision Systems GmbH and David 3D Solutions.

As more manufacturers embrace the technology and install the equipment in their factories, there will be a shift in where the producers of 3D printers obtain their revenues. As has been the case in other sectors, aftermarket services for 3D printers will be where the money is, through materials, maintenance, and repairs. In many sectors employing that model, 80% of revenues are generated through operational expenditures, not capital, and there is no reason to think that 3D printing will be any different. However, there is still plenty of potential for 3D printer sales, and hardware sales will continue to dominate in the shorter term. According to Wohlers, there has been a growing demand for 3D printers for many years, and 2017 saw a 21% increase in 3D printer sales overall, with metal 3D printers rising by 80% that year.[10]

Beyond industrial 3D printing, it is worth a short observation about consumer 3D printing, the idea the technology will see developments like those witnessed by to the PC and 2D printer, and that we'll have 3D printers in our homes. As mentioned in the Introduction, this is not covered in this book. However, much as the development of the PC and 2D printers for home use drove changes in the capabilities, sales, and prices of those items, it is quite conceivable that if a company were to enter the home market, the 3D printing sector would undergo a similar change. The challenge here is that consumers are impatient, wanting things when they need them, not hours later. They are largely unskilled, so using CAD is not an option. They want something reliable with at least some level of quality assurance, and 3D printing is not yet able to provide those with the lower prices that home use would need. Last, the narrower materials capabilities of 3D printers for the foreseeable future means that there won't be the flexibility in what can be produced to make 3D printing something for the home. Were those dynamics to change, however, then 3D printing would certainly be more pervasive everywhere.

The development of 3D printing technologies has long been shaped by a pursuit to resolve its many constraints, from cost and accuracy to material properties and production speeds. However, as 3D printing continues its transformation from a tool for rapid prototyping to industrial production, there is an increased realization that these factors need to be traded off, that particular 3D printing technologies can be suitable for some of those, but that no one technology—much less a single 3D printer—will be able to resolve all the challenges of speed, cost, precision, material properties, and part size at the same time. This will inevitably lead makers of 3D printers to diversify their offerings, employing developments in technology and materials to give supply chains a choice of solution to meet their particular needs. It is likely to also see a fragmentation of the 3D printer sector, with new entrants focusing on a small configuration of those offerings. For instance, the firm Apis Cor, founded in Russia in 2014 by Nikita Chen-jun-tai, provides mobile-construction 3D printers specifically for producing whole buildings. The long-established German engineering company BEGO GmbH & Co. produces the Varseo S 3D printer specifically for applications in dentistry, using stereolithography to make casts, guides, and temporary dental fixtures. New uses are also driving those developments: for instance, the desire to be able to 3D-print fully working, multimaterial objects with embedded electronics is driving developments in 3D printing processes to produce the necessary precision and tolerances as well as materials, with some researchers successfully using carbon nanotubes to achieve those.[11]

WHERE NEXT?

In the short term, the sectors that will see most penetration from 3D printing are the transport sectors, principally aerospace and automotive, and healthcare, both medical and dental. These sectors have been using 3D printing for decades. The operating models of the transport sector involve a high number of parts that need to support long-lived platforms. Those parts often need some adjustment to fit the specific platform and, being long-lived, that leads to a high risk of obsolescence, a high cost of making parts in lots of one, and a high cost to store all that inventory. Aerospace companies currently use 3D printing to accelerate the manufacture of parts from GE's fuel nozzles to back-of-seat tray brackets, and that is expanding to other components from air ducts to support struts and even jet engine blades. The automotive sector, the earliest adopter of 3D printing, already uses the technology to produce new designs and prototypes; for example, the Ford Motor Company produced its millionth 3D-printed component in 2015. Beyond that, the automotive sector is moving toward incorporating 3D printing in its aftersales market, providing spare parts for service centers (something many aerospace firms have ambitions of doing) and, potentially, directly to customers. The vision is that customers will soon no longer have to go into an auto service center and wait for two or three days for a part to be shipped across the country or even imported; rather, the service center will be able to produce the spare part to order, either at its premises or at a nearby 3D printing center. Mercedes-Benz Trucks announced that it will be providing a 3D-printed spare parts service, following a similar initiative at Audi, starting with 30 parts for the Actros series of trucks in September 2016. Other sectors with similar characteristics are following close behind, such as oil and gas, mining, and defense—capital-intensive sectors where equipment has a very long lifecycle, typically decades.

In the healthcare sectors, the capabilities of 3D printing to produce items that are customized to the patient (e.g. training models, implants, or supports), and to do so competitively, give it a tangible benefit. The advent of bio-printing also offers exciting possibilities, promising to revolutionize medicine. For instance, Erik Gatenholm, founder of bio-printing firm Cellink, said in 2017 that his goal has always been:

> "To change the world of medicine—it was as simple as that. And our idea [is] to place our technology in every single lab around the world."[12]

Being able to reduce the number of times a patient visits the dentist for an implant or to adjust braces results in happier patients and more productive dentists. The greater freedom of location that 3D printing offers opens up the opportunity for its use in humanitarian situations, providing mass healthcare where this wasn't previously possible.

Naturally, the significant role that 3D printing plays in rapid prototyping will continue to expand. Even the fast-moving consumer goods sector (FMCG) already uses the technology for the production of prototypes of new packaging, for example. As those companies have found, the advantages of fast redesign and production mean that market research will increasingly make use of 3D printing to give customers the look and feel of potential new designs. Iterating those designs with customer feedback will be carried out far more quickly and cheaply than has ever been possible (see Chapter 6).

However, it is in mainstay manufacture that 3D printing has the biggest potential to significantly redraw supply chain operating models, especially in those sectors where there is a need for personalization, whether due to the purpose of the object or the desire of customers. In sectors such as jewelry, fashion, and consumer goods, the availability of unique designs is a part of customers' wants, and 3D printing can provide for that via its "lot of one" capability, with little or no increase in the cost per item than for higher-volume production runs, provided the added design burdens can be overcome. Several designers in those sectors are using it, including jewelers Chanel and Guy & Max, and sports shoe maker Adidas, and many others are considering it to expand the number of variations to their products, from having more colors to customizing decorations, with little increase in cost and little impact on time to manufacture. Moreover, those designs can be evolved and changed rapidly with little time between cycles. Taken to its extreme, this opens the possibility of jewelry and fashion products that are bespoke and customized to the individual person. In 2014, Avi Reichental, the former President and CEO of 3D Systems, said in a TED talk:

> "You all know your shoe size. How many of you know the size of the bridge of your nose, or the distance between your temples? […] Wouldn't it be awesome if you could […] get eyewear that fits you perfectly and doesn't require any hinge assembly, so chances are the hinges are not going to break?"[13]

It isn't only in manufacturing goods directly that 3D printing is making headway. The use of 3D printers to make casts and molds for high-volume manufacture is an accelerating trend, one already bringing huge benefits. By using the technology, the mold and cast designs can be iterated faster to arrive at the final form more quickly—and hence more cheaply—than using traditional approaches. The manufacturer Seuffer, which makes parts for household appliances and commercial vehicles, has been using 3D printers for over three years to produce injection molds. They have achieved a 97% reduction in the cost to produce each mold, and the time to make one has fallen from eight weeks to a few days, including design, which allows them to iterate those designs several times without previous constraints on cost and time. The performance of those parts, in terms of the pressures and temperatures that the molds work with, has been commensurate with those that were traditionally machined.

3D printing is already a more capable set of technologies than many of those working in supply chains realize. With the benefits that it brings and the future course that the technologies and industry dynamics are indicating, understanding how to access 3D-printing capabilities and what the different options for that are sets the scene for describing how 3D printing will drive supply chains.

Notes

1. Hayley Lind, "Volkswagen Sees a Future in 3D Printing Car Parts," The Drive, November 24, 2017, http://www.thedrive.com/sheetmetal/16374/volkswagen-sees-a-future-in-3d-printing-car-parts.
2. Roopinder Tara, "A First Look at Figure 4, Industrial 3D Printing from 3D Systems," Engineering.com, September 20, 2016, https://www.engineering.com/3DPrinting/3DPrintingArticles/ArticleID/13155/A-First-Look-at-Figure-4-Industrial-3D-Printing-from-3D-Systems.aspx.

3. Vance H. Trimble, *Sam Walton: The Inside Story of America's Richest Man* (New York: Dutton, 1990).

4. Bernard Marr, "Big Data: 20 Mind-Boggling Facts Everyone Must Read," *Forbes*, September 30, 2015, https://www.forbes.com/sites/bernardmarr/2015/09/30/big-data-20-mind-boggling-facts-everyone-must-read.

5. Watson Customer Engagement, "10 Key Marketing Trends for 2017 and Ideas for Exceeding Customer Expectations," IBM Marketing Cloud, July 18, 2017, https://www-01.ibm.com/common/ssi/cgi-bin/ssialias?htmlfid=WRL12345USEN.

6. Acatech, "Recommendations for Implementing the Strategic Initiative INDUSTRIE 4.0: Final Report of the Industrie 4.0 Working Group," April 2013, https://www.acatech.de/Publikation/recommendations-for-implementing-the-strategic-initiative-industrie-4-0-final-report-of-the-industrie-4-0-working-group/.

7. Jeremy Reimer, "Total Share: 30 Years of Personal Computer Market Share Figures," Ars Technica, December 15, 2005, https://arstechnica.com/features/2005/12/total-share/6/.

8. Steve F. A. Acquah, et al., "Carbon Nanotubes and Graphene as Additives in 3D Printing," in *Carbon Nanotubes: Current Progress of their Polymer Composites*, ed. Mohamed Berber and Inas Hazzaa Hafez (London: IntechOpen, 2016, July 20): 227–51.

9. Wohlers Associates, *Wohlers Report 2017*, http://wohlersassociates.com/2017report.htm.

10. Wohlers Associates, *Wohlers Report 2018*, http://wohlersassociates.com/2018report.htm.

11. Jung Hyun Kim et al., "Three-Dimensional Printing of Highly Conductive Carbon Nanotube Microarchitectures with Fluid Ink," *ACS Nano* 10, no. 9 (2016): 8879–87, doi: 10.1021/acsnano.6b04771.

12. Maddy Savage, "The Firm That Can 3D Print Human Body Parts," BBC, November 15, 2017, http://www.bbc.co.uk/news/business-41859942.

13. Avi Reichental, "What's Next in 3D Printing," Tiny TED, https://en.tiny.ted.com/talks/avi_reichental_what_s_next_in_3d_printing.

CHAPTER FOUR

4

Accessing 3D Printing Capabilities

N THE 2007 TV series *Mad Men* about the growth and evolution of a 1960s advertising agency, there is an episode centered around the arrival of the company's first business Xerox photocopier. As depicted in the show, this was a significant moment for any company at the time, when it could begin to do work in-house, freeing itself from the need to send out copying jobs to a third party or use the prevalent, solvent-rich Gestetner copying (aka ditto) machines. The early business photocopiers, which had first gone on the market in the 1950s, were large, cumbersome and, above all, required a significant capital investment. In the 1960s and 1970s, sales volumes in photocopying companies like Xerox grew steadily—indeed, that company's brand became the byword for photocopying itself. Over that time, clients bought increasingly more photocopiers, which also became more capable and more expensive. However, toward the end of the 1970s, sales began to wane. Why?

Businesses that had acquired their own photocopiers had essentially overspent—there was more capability, more capacity, than they needed in their machines. The result was that they didn't need to continually upgrade, and thus they cut back on spending, leaving photocopier manufacturers contemplating how to restore their previous rising revenues with new approaches. They came up with a solution in the 1980s: no longer would they rely on sales of large-scale photocopiers. Instead, they would retain the machines themselves and lease them to corporations under a service agreement. All upgrades, maintenance, and repairs would be covered by the agreement and the customer would be charged on the basis of the volume of work they used the photocopier for. This change opened new operating models, granting small- and medium-sized clients access to the same sorts of services as large firms. Not only did the sector improve its sales; the new operating model was subsequently adopted in other sectors.[1]

Much as for photocopying, a company can access 3D printing capabilities via three models:

1. Acquiring 3D printers
2. Leasing 3D printers
3. Employing 3D printer hubs

This chapter will examine those in turn.

 ## ACQUIRING 3D PRINTERS

In the mid-1980s, after the patenting of the stereolithography process and the founding of 3D Systems, the first company to commercialize the technology, Ford Motor Company bought the third 3D printer ever sold. Ford saw the potential for the technology to help it design new automotive components and vehicle prototypes. Acquiring a 3D printer in the early days, as with any new technology, brought with it the cachet of simply owning such a futuristic technology. By the 1990s, companies would have to spend hundreds of thousands of dollars to buy a high-quality 3D printer, even more if metal materials were needed.

The early 2000s began to see a change. As the original patents expired and new entrants to the market emerged, prices began to fall. New materials could be used in 3D printing with ever-higher levels of accuracy and precision. Although average printer prices declined, particularly at the lower end of sophistication, high-spec machines that can make objects in specialized materials or with the highest levels of accuracy and precision continue to be expensive. Additionally, those machines, as with any such capital items, need regular maintenance and repairs, both of which can be expensive in cost and lost production time. Today, there is a plethora of 3D printer manufacturers, with the market by revenue dominated by GE, Stratasys, 3D Systems, and EOS, which among them sell over 50% of all machines. Specialist firms have emerged steadily over the last 10 years, each targeting a particular industrial need or material, such as Renishaw in the UK, Carbon3D in the USA, and Voxeljet in Germany. The sales center of gravity saw a significant shift in 2016: more units were sold that year in China than the USA.[2]

Acquiring a 3D printing system is very much like acquiring any complex equipment, with the same sorts of contracts and payment terms. However, acquisition is a significant investment and it is not the preferred first step for most companies.

LEASING 3D PRINTERS

The lease model that Xerox led in the photocopier industry has since been adopted by other sectors. In fact, a similar approach has been used in aerospace through Rolls-Royce's so-called Power-by-the-Hour model since 1962. Through it, the airlines that operate aircraft fitted with Rolls-Royce engines don't own those engines themselves: instead, Rolls-Royce charges for them on the basis of a fixed cost-per-hour that they are used for. Maintenance, repairs, and upgrades are all included in the fee and taken care of by Rolls-Royce, rather than being burdens that the airline must pay for separately. The risk of owning the engines is therefore transferred from the airline to the manufacturer, with the former not having a need to concern itself with the complexity and difficulty of owning a large capital item—the engines—and its upkeep.

The events that transpired in the photocopying sector in the 1980s are a close match to recent developments in the 3D printing sector. As the hype about 3D printing rose, so did sales of 3D printers, and the share prices of their makers shot up. When their customers found that the machines weren't as capable as they thought they would be and became disillusioned with what they could do, sales dropped and share prices plummeted. The 3D printing sector quickly learned the lessons of the past; in addition to pushing to develop better technologies and better capabilities, they changed the way that companies could access 3D printers, offering leasing arrangements instead of direct purchase. Already, 3D Systems, EOS, and Stratasys are offering similar service-based leasing contracts to their customers. This model brings benefits to both the user and the manufacturer. For the former, there is no large capital outlay, with expensive through-life maintenance and repairs that must be provided for in-house or via a specialist contractor. For the latter, there is a longer-term contract with the customer, a steady revenue stream, and a more manageable portfolio for the planning of maintenance, repairs, and upgrades.

Leasing agreements for 3D printing systems are similar to those for other major equipment. They will cover the installation, activation, and ongoing support that the customer can expect (including access to maintenance and repair services). Often, they will define the terms of use of the equipment, describing how it can be employed and what for.

 ## PRINTER HUBS

When the photocopier providers changed their sales models, third-party photocopying companies began to arrive, able to invest in obtaining the capital equipment, then providing photocopying services on demand to organizations and individuals. Names like Office Mart, Staples, and Prontaprint began to offer business-quality photocopying services to general audiences—reverting business models to the pre-Xerox era of outsourced copying. Until the 1990s, when students wanted to give their handwritten or hand-typed thesis a professional look, they would send it to a local printer who would give it a cover and bind it. Later, when that thesis was rendered digitally on a word processor, that document would be taken to the printers on a floppy disk to be produced on the good-quality professional machines, which were too big and expensive to be owned by the student or even the university. Businesses similarly would use those printer hubs for large-volume orders that went beyond their own internal capabilities and budgets. Something similar has been taking place in the 3D printing world, with the emergence of 3D printer hubs. These companies bring the same benefits as those traditional hubs, giving businesses and individuals access to 3D printing capabilities that they otherwise would not be able to afford or not want to own themselves. They typically contain a range of 3D printers that satisfy most of their customers' technological needs, and increasingly can provide higher volumes of productions.

The great advantage of 3D printer hubs is that organizations can try out 3D printing and new technologies, produce low volumes, and outsource their manufacturing completely, without the need for the large investments to acquire or lease a large machine, postponing that decision until they are convinced of the business case. Moreover, 3D printing hubs provide a means of quickly accessing scale to deal with changing or spiking demand, thus increasing the responsiveness of a supply chain.

This hub model is also adapting to the needs and dynamics of the market. For instance, to remain viable and economic, the hubs must maximize the amount of time that the machines are working. Companies like 3Discovered and 3D Hubs are now giving customers visibility of their machine time so that they can find available manufacturing capacity when and where they need it. Much as traditional production does by using Manufacturing Resource Planning (MRP) systems, this allows those hubs to maximize their printers' utilization. This also favors customers, who then have access to as much manufacturing capacity as they need (within the limits of 3D printing, of course). By 2017, those hub-based 3D printing providers, which industry journalist T. J. McCue calls "Ubers of 3D Printers,"[3] had arranged for the production of over 600,000 parts for various customers across a network of over 2,000 printers in 160 countries. 3D Hubs states that this model means that "one billion people on this planet already have a 3D printer within 10 miles of their homes."[4]

Many distributors of 3D printers have been adding printing services to their portfolio. Objective3D, an Australian firm based in Carrum Downs, began life providing 3D printers to the Australia and New Zealand markets. In 2014, the company opened the doors to its 3D printing manufacturing plant in Melbourne, offering companies a Manufacturing-as-a-Service (MaaS) 3D printing capability, with a capacity to produce 150,000 items annually.[5] They aimed their proposition particularly at clients in the mining, defense, automotive construction, and biomedical sectors, which use their facilities to pilot their own adoption of 3D printing, prior to acquiring or leasing in-house capability. Objective3D has enabled its clients to make items across the full end-to-end production development life, from prototypes through jigs and fixtures, to end-use, low-volume manufactured parts.[6]

Strategically, the 3D printer hub model has thrown a lifeline to the logistics and 2D printer hub sectors, opening up new business models that leverage 3D printing's capabilities and extend their relevance in the face of changing customer demands. The company Fast Radius operates such a hub model in a symbiotic relationship with UPS; together they are able to produce finished products in a broad range of materials, from exotic metal alloys to ceramics in a true MaaS approach (see sidebar). Well-known print shops like Staples have begun to trial the printer hub concept, giving their customers access to 3D printers at their outlets in the USA and UK.

Manufacturing-as-a-Service

The development of 3D printer hubs has led to the realization of the MaaS concept. Much as other "as-a-service" models involve networks to provide value, so MaaS enables companies to use cloud-based, networked manufacturing to produce items on demand. This can be done to minimize lead times, such as by employing MaaS infrastructure closer to customers, as well as to provide scale, such as via the Fast Radius service models. This capability remains available without the MaaS user having to invest in the acquisition, setting to work, maintenance, repair, or upgrading of the 3D printing infrastructure, or developing the skills to use the equipment and software. Instead, MaaS users can rely on the specialist skills and expertise of their providers, itself developed by accelerating and accumulating experience in the operation and optimization of 3D printing.

 MAKING THE CHOICE

Several factors will inform which of these three options is the best for a supply chain. Those include:

- How developed the idea of embracing 3D printing is
- The type of 3D printer required
- The volumes of items that will be 3D printed, both in the short and longer terms
- The available budget
- The amount of control over the 3D printing process that the supply chain wants to retain

If a supply chain is seeking to test what 3D printing can do, then using a 3D printing hub is preferable. If it wishes to retain complete control over the 3D printing process, as may be needed for making end-use items for the aerospace sector, then acquiring a 3D printer or a more sophisticated leasing arrangement is preferable. Similarly, if budgets are restricted, then outsourcing 3D printing has the benefit of minimizing expenditure. The combinations of just these five factors are myriad and each supply chain will have its own preferences and risk appetites.

The first part of this book has taken a closer look at the technical elements of 3D printing, from what it is to the types of technology that 3D printing can involve. It has looked at the characteristics of 3D printing and made an informed prognosis of where 3D printing is heading, before describing the ways that 3D printing capability can be acquired, and these have all been done to provide those involved in supply chain decisions with the grounding to understand the terminology and technical issues. The second half of this book now focuses harder on the impact 3D printing has on the supply chain, starting with a case example that describes how 3D printing can drive supply chains.

 Notes

1. John Hauer, "What Really Happened with Print: History Lessons for 3D Printing Equipment Manufacturers," 3dprint.com, May 18, 2015, https://3dprint.com/66002/3d-printing-manufacturers/.
2. Nick Hall, "3D Printing Explodes in China," 3dprintingindustry.com, May 21, 2016, https://3dprintingindustry.com/news/3d-printing-explodes-china-79707/.
3. T. J. McCue, "3D Printing Is Changing the Way We Think," *Harvard Business Review*, July 21, 2015, https://hbr.org/2015/07/3d-printing-is-changing-the-way-we-think.
4. 3D Hubs, accessed February 18, 2016, www.3dhubs.com.
5. Objective3D Direct Manufacturing, accessed January 10, 2018, www.direct3dprinting.com.au.
6. Babs McHugh, "Australia's First Commercial 3D Printing Factory Opens in Melbourne," ABC News, October 12, 2014, http://www.abc.net.au/news/rural/2014-10-13/3d-printing-seen-as-manufacturing-route-for-the-future/5807984.

CHAPTER FIVE

Deutsche Bahn: Applying 3D Printing to the Supply Chain

THE GERMAN RAIL company Deutsche Bahn AG has a hugely complex supply chain. The largest railway company in the world by revenue, it employs over 300,000 people across 130 countries. It owns, operates, maintains, and repairs some 24,000 units in fleets of local, national, and international trains across Europe and beyond, as well as a wide network of stations, tracks, and other infrastructure, carrying passengers on over four billion journeys each year and annually moving over 300 million tons of cargo. It has to deal with assets that are based on both old technologies, such as the rail network, as well as new developments, like the European Train Control System (ETCS)—the continent's newest signaling and control infrastructure.

This broad portfolio of assets is capital-intensive, requiring the investment of billions of euros on trains that have lifecycles measured in decades; the latest models of locomotives and carriages in its fleet are over 40 years old. In the past, Deutsche Bahn and its predecessors in West and East Germany had their own internal manufacturing departments to make spare parts for their fleets. However, late in the twentieth century, cost pressures led Deutsche Bahn to cease this manufacture and instead to rely on suppliers alone, either by buying the likely inventory of spare parts at the beginning of a contract or later in the lifecycle of the equipment. The combination of a large and long-lifed fleet with the need to have parts available when required led to huge inventory; today Deutsche Bahn typically hold dozens of millions of spare parts from a list of over 100,000 SKUs, all stored in many distribution centers, warehouses, and workshops across Europe and beyond.

The company spends some €600 million on spare parts each year just for their rolling stock and a reduction of even 1 to 2% would be a considerable savings. As well as locking up working capital, that inventory incurs significant storage, insurance, maintenance, and management costs. Many of the parts are slow to be used, only required once every few months or longer, and some might never be used at all. This exposes Deutsche Bahn to a huge risk of obsolescence, carrying parts for decades that it will never need. Moreover, when some of those items are fully consumed and orders need to be placed to restock them, the original suppliers may well have switched their manufacturing operations and no longer offer the original parts; others may have gone out of business altogether.

In those situations, the company has to work with its suppliers to have them retool and reset manufacturing lines to make those needed parts, or find new suppliers to produce them. Both options are expensive and prolonged, resulting in lead times of many months—18 months is not unknown. The result is that Deutsche Bahn's operations are very cash intensive, and service levels, in terms of trains being operational and available, are lower than they could be. Ultimately, customers suffer from higher ticket prices and longer and more frequent delays. In some of the locations where the company operates, that lower service level exposes the company to fines and possible loss of their operating license.

A BETTER WAY OF DOING THINGS

It was in the face of these pressures that a team of researchers and engineers considered an alternative approach to spare parts manufacture and management: 3D printing. With members of the team understanding what 3D printing could do, and the constraints that the technology presents, they began a campaign to pilot its use for making some of the needed spares faster and more cheaply. Initially, they assessed that the biggest opportunities lay in fabricating parts for end use; as Deutsche Bahn didn't develop parts, there was no need for prototyping. Moreover, although tooling was obviously used, the volumes of required tools (e.g. jigs, fixtures, and other such equipment) were relatively insignificant. Spare parts, however, numbered in the many thousands, often in small batches or single units. Thus, in October 2015, they began to work on those.

The first step was to identify which of the thousands of spares for their trains, stations, and associated infrastructure could and should be 3D printed. To do that, a small analytics team was tasked with collating the materials data that Deutsche Bahn held, together with data from suppliers who were able and willing to share it, to provide the necessary geometric, materials, and other technical information needed. After three months, the team reported that this was a fruitless task, that the necessary technical data simply was not meaningfully available. Part of the challenge lay in how trains had been procured: Deutsche Bahn had abandoned its own manufacturing capabilities years previously, and it had become common for design data to be excluded from contracts with suppliers. Another approach was needed.

The 3D printing team instead decided to take their message to those who understood the spare parts best: the engineers and technicians in the many workshops who maintained and repaired the trains. They prepared a set of presentations on what 3D printing is, what it can and can't do, what materials can be used, and the advantages and drawbacks of the technologies. This included conveying the message that the number of materials available are restricted in terms of range and the number that can be used in a single part, restrictions in size, and so on. They then embarked on a tour of the main workshops, giving the presentations and then asking the technicians to identify which items they would advocate being piloted.

The first items identified were the simple plastic coat hooks that are found in passenger compartments. In the older train models, the hooks were no longer available from the original supplier. Discussions with potential suppliers found that the costs of tooling up and injection-molding a new set of hooks were in the region of €60,000, with the hooks costing a few cents each thereafter. In comparison, discussions with a specialist computer aided design (CAD) designer and 3D printing bureau found the initial cost of design work to be €100, and then producing the hooks with an selective laser sintering (SLS) machine would cost €10 each.

The technicians also identified their first metal part: a terminal box that sat at the front of the model blocks on ICE1 trains (Figure 5.1). This 2.5 kg item is normally made from molded aluminum and typically requires replacement up to five times each year on a single engine, due to damage from stones being kicked up along the tracks when the trains are operational, and occasionally from damage incurred when the engine was placed on workbenches. The lead times of replacements were about one year. To shorten this delay, the

FIGURE 5.1 ICE1 terminal box (small box on left side of the engine) (By kind permission of Deutsche Bahn/Siemens).

team decided to use 3D printing, reverse-engineering the design and producing a unit in four weeks, thus mitigating the sidelining of a train, which would cost tens of thousands of euros per day. Another item, a steel sandbox, which stores the sand that is spread in front of a train's wheels to give it grip, would have taken 15 to 18 months to engineer and produce as a replacement (Figure 5.2). The team produced a replacement in three weeks. Moreover, the new item was

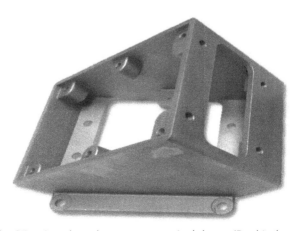

FIGURE 5.2 3D-printed replacement terminal box. (By kind permission of Deutsche Bahn/Siemens).

made in titanium to give it extra strength, something that would be been inconceivable without 3D printing technology, and only enabled because the savings in machining costs from using SLS/selective laser melting (SLM) outweighed the increased costs of using the stronger material powders.

Consulting the workshop personnel also brought with it the benefit of identifying parts for 3D printing that a simple data analysis would have missed. One item that the technicians had to replace on a regular basis was a complicated, spring-loaded locker assembly. This part required replacement anywhere between 10 and 40 times a year, and each repair would take up to over an hour. The entire assembly (Figure 5.3) was the lowest replaceable part available from the supplier at a cost of €100 each time. The technicians had found from experience that the item typically needed replacement because of the failure of a small, plastic grommet within it (Figure 5.4), a sleeve that provided some protection at one of the mounting points and cost €1 each, but that was not available from the supplier as a separate item. The team then took one of the parts, produced a CAD file of its dimensions, identified the right type of plastic to make it in, and 3D printed them. Doing so cut the repair time to five minutes and saved €10,000 per year.

Deutsche Bahn also found benefits in the area of manufacturing sustainability. A spare aluminum headrest that was installed in many of the trains would traditionally be made by milling a large block of that material. Instead, the team opted to 3D print the replacements (Figure 5.5), with an end-to-end production time of two weeks and a drastic reduction in material waste.

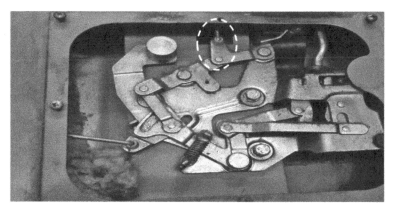

FIGURE 5.3 Locker assembly, with the grommet indicated (By kind permission of Deutsche Bahn/Siemens).

FIGURE 5.4   Grommet in the locker assembly (By kind permission of Deutsche Bahn/Siemens).

The freedom of design and the benefit of being able to produce lots of one has also had an unexpected benefit for an often ill-served group of Deutsche Bahn passengers: the sight-impaired. These passengers rely on signs being available in Braille, something expensive to do en masse as each sign needs to be constructed individually and frequently uniquely. The 3D printing team is

FIGURE 5.5   3D-printed replacement headrests (By kind permission of Deutsche Bahn/Siemens).

FIGURE 5.6 3D-printed Braille signage (By kind permission of Deutsche Bahn/ Siemens).

now providing their capabilities to station managers who can identify useful and convenient signage, from station names to directions (Figure 5.6). They pass these requirements to the team, which commissions the signs to be made in aluminum in a short few days, and the finished products are then simply and quickly installed at the station.

ADAPTING SUPPLY CHAINS

In its first year of operation, the 3D printing team only succeeded in producing 10 parts, restricted by the lack of engagement and belief in their approach in the wider engineering and operational areas of Deutsche Bahn. Having proven that there are cost and time—and therefore service level—benefits to using 3D printing, the company is now embedding the technology into its operations and adapting its spare parts supply chain accordingly. By March 2018, the company had 3D-printed 4,500 parts, and it was quickly ramping up to over 15,000—early assessments of its inventory found that current 3D printing technologies could be used for 10 to 15% of all SKUs. The company is part of a wider industry group, Mobility Goes Additive, that brings together highly experienced and capable 3D printer makers, software engineers, designers, and operations specialists with the goal of advancing the application of 3D printing. Participation is continuously growing, with new suppliers, service providers, and transport companies joining, and the group examines how the technology can be used to produce items with the right quality, how to develop standards for using the technology in the transport sector, and how to disseminate knowledge of what 3D printing can do.

Internally, there is a growing recognition in the company that using 3D printing is not a technology issue, but rather a change management topic, and needs to be responded to accordingly. The first step of this response is the use of data analytics to support the identification of further parts for 3D printing, using a combination of bills of materials, procurement lead times, maintenance and design data, and a thorough modeling of the total cost of ownership of parts. Once a candidate part has been identified, a legal team looks at its intellectual property (IP) situation to understand any IP issues (parts whose designs are over 25 years old are no longer covered by any original patents). The team also looks at regulatory issues; many parts are covered by EU safety legislation because they are destined for transport use. For instance, flame retardancy requirements are very high and inform what materials can be used. Discussions with the original suppliers—where those still exist—then conclude how 3D printing can be used to provide the workshops with the necessary parts, such as using a printer bureau to make the parts for the supplier, who then resells them to Deutsche Bahn, or by Deutsche Bahn directly employing that bureau. The company has already established partnerships with software companies like 3YOURMIND, printer bureaus and 3D printer makers like BigRep, having calculated quite early that using such outsourced providers was cheaper than buying their own printers and retaining their own in-house expertise. With volumes ramping up, the company is now adjusting its supply chain to be proactive, using predictive demand management to set requirements for coming years and tendering the production of groups of parts to 3D printing bureaus and companies.

Deutsche Bahn's procurement teams have responded to the prospects for expanded 3D printing of parts, and their processes are increasingly aligned, optimized, and centralized. Design data, covering geometries, materials, and other technical matters, is now routinely included in the package when new trains and equipment are acquired. The supplier base is also adapting to the new dynamics. Concerns about liability for parts that the printer bureaus produced for the company initially resulted in the inclusion of a clear statement in supply contracts that the parts were provided for prototyping purposes only, not end use. This naturally restricted how Deutsche Bahn could use those parts as it affected their own exposure to risk. Following a series of discussions and agreements on who is responsible for the testing and quality assurance of fabricated parts, the standard approach is for suppliers and bureaus being responsible for ensuring that what is fabricated is what was stipulated in the original requirements, while final quality assurance is provided by Deutsche Bahn, as it would do however of how the parts are made.

More widely, there is now a more widespread understanding of the capabilities of 3D printing and how it changes supply chains. With results yielding cost savings and reduced lead times, the evidence is clear. As Uwe Fresenborg, CEO of Deutsche Bahn Fahrzeuginstandhaltung (Vehicle Maintenance), told 3dprintingindustry.com:

> "For the maintenance of our vehicles, we need immediately available spare parts. Our trains are expected to roll [and] 3D printing helps us in doing so. Printing is faster, more flexible and cheaper than conventional manufacturing processes, and the vehicles are available again in a very short time and are used for our customers."[1]

The Deutsche Bahn case example demonstrates practically many of the ways that 3D printing drives supply chains. Understanding the impact of each of the supply chain elements—Plan, Source, Make, Deliver, Return, and Enable—will enable those who make decisions about them to grasp how the technology can impact those too.

Note

1. C. Clarke, "Deutsche Bahn Extends Use of 3D Printing to 'Revolutionize Maintenance.'" 3dprintingindustry.com, May 15, 2017, https://3dprinting industry.com/news/deutsche-bahn-extends-use-3d-printing-revolutionize-maintenance-113320.

CHAPTER SIX

The Impact of 3D Printing on the Supply Chain

TODAY'S SUPPLY CHAINS are facing a fast-changing landscape of economic and demand pressures. The world of Henry Ford, with its mass production of things that came in "any color, so long as it's black," has all but disappeared into the mists of time. Other than continuing to want the best price possible, today's customers are very different from those of the past:

- They want more customization and personalization from their products.
- They have a far lower tolerance for waiting, be it for new designs or things to arrive at their doorstep or operations.
- They change what they want faster than ever, so product lifecycles are shorter than they used to be.
- They are more likely to compare offerings and negotiate, looking to increase the value of what they get.

Together these result in more complex demand forecasts, which can quickly balloon into ever more bloated inventories and higher obsolescence costs. Supply chains need to be more responsive and agile, which has led to the emergence of several solutions, techniques, and tools. Today's supply chains increasingly use automation, from robots in manufacture to robotic process "bots" and intelligent agents, to increase productivity, reduce work adding no value, and lowering error rates. They employ data analytics to identify issues before they become problems, to optimize operations, and to offer their customers an advantage where even the most marginal benefit is sought. As supply chains became more complex, having to cater to the needs of "omnichannels" with elements spread across wide geographies, their management became exponentially harder, with more risks threatening them. There is now a greater need for strategies and actions that mitigate risks and, wherever possible, eliminate them, such as by seeking ways to shorten long supply chains, to become more responsive and flexible, and to develop ways to deliver precisely what the customer wants, where they want it, when they want it.

Over the last 60 years, companies have succeeded in managing those pressures by using platform-based design and postponement, using common core designs and delaying final additions and alterations to meet customer needs further down the value chain, closer to those customers. The automotive industry started this with chassis platforms and engines which were then productized into individual models. For instance, the recently launched Jaguar XE is a saloon car with the same chassis and engine as the sporty coupé F-Type. That same methodology was notably used by the mobile phone handset manufacturer Nokia, which in the 1990s and early 2000s moved from an operating approach of designing and manufacturing individual models separately to creating a small set of common platforms from which it could respond to customer needs, producing new models so quickly that its product pipeline became known as the "Nokia machine gun." In today's environment, maintaining the manufacturing capability and capacity to accommodate the nuanced demand from customers increases costs, eventually becoming unsustainable. New solutions to manage the supply chain are required, and 3D printing has shown itself to be one.

3D printing impacts all parts of the supply chain, not just the usually discussed area of manufacture. Understanding its impact is critical to deciding whether to use it. This chapter will take a close look at how 3D printing affects each of the component parts of the supply chain, examining what benefits and challenges it brings to each. This will include an examination of the key supply chain metrics that are impacted, building on the points already made in earlier chapters.

DEFINING THE SUPPLY CHAIN

To fully appreciate how 3D printing affects the supply chain, we must agree on what the supply chain is. All too often, "supply chain" is interpreted to be the movement or storage of things (i.e. logistics). At other times, it is considered to refer to the sourcing and procurement of materials and services. It is much more than this, encompassing many parts of a business, and acting as the backbone to operations. While there are plenty of models that define what a supply chain is, the one most widely accepted and understood is the Supply Chain Operation Reference model (SCOR). Developed in the mid-1990s by the Supply Chain Council and updated regularly since, SCOR describes the supply chain in five key business processes:

- Plan
- Source
- Make
- Deliver
- Return

These are underlined by a sixth process: Enable. Each of these processes contains the subprocesses and activities, metrics, and best practices relevant to that part of the operating model. As Shoshannah Cohen and Joseph Roussel say in their authoritative book, *Strategic Supply Management*:

> "The SCOR® model provides a framework and standardized terminology to help organizations integrate a number of management tools, such as business process re-engineering, benchmarking, and best-practice analysis. The SCOR® toolbox enables organizations to develop and manage effective supply chain architectures."[1]

In this architecture, the six processes cover the business thus:

- *Plan.* The set of processes related to getting the supply chain ready for operation, from demand planning to what will be supplied to meet that demand and how that will be achieved.
- *Source.* Those processes related to the acquisition of materials and services, from identifying sources to the actual procurement processes itself.
- *Make.* The manufacturing processes, where goods are made, and services developed, for onward sale to satisfy customer requirements.

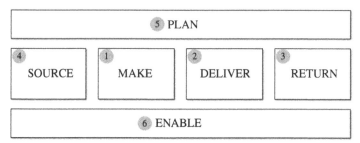

FIGURE 6.1 The supply chain model. Numbers indicate order of discussion in chapter.

- *Deliver.* The storage and movement of materials, work in progress (WIP), and finished goods, before, during, and after manufacture.
- *Return.* The reverse flow of materials, typically from customers back to the company, such as for damage-related, end-of-life, or recycling purposes.
- *Enable.* All those activities that support the operation of the supply chain, from data and information flow, to finance, marketing, and general management and administration.

Of course, different businesses have different forms and supply chains will naturally vary. The strength of the SCOR model is that it is a strong architecture that can be used to understand virtually any business and, more relevantly, to see where and how different factors will affect supply chains, such as the effect of 3D printing.

To properly and more easily appreciate how 3D printing impacts each part, this chapter will look at them in the order indicated by the numbers in Figure 6.1.

3D PRINTING AND "MAKE"

Today, the most evident advantages of the technology are those related to making things, both in terms of cost and time. 3D printing also allows for true demand-driven manufacturing, where production is discrete and often customized, based on an actual order or per consumption instead of relying on forecasts. This is the case whether that demand arises externally and internally, from outside customers or from the internal supply chain needs.

Moreover, the end customer can now be involved at every stage, in every step of the production process, from conceptualization to finished product, and that in itself has a transformational effect on manufacturing: with that collaboration, both users and producers become designers, testers, and assessors.

As described in Chapter 2, 3D printing an item is itself a slow process compared to traditional manufacturing techniques, taking hours if not days rather than seconds or minutes. When the entire time to make the part is considered, however, from tooling up and setting machinery to work to producing the part itself, 3D printing is clearly a much faster option. Moreover, the advantage of 3D printing to make items with fewer individual components, each of which might otherwise need its own production line, also results in fewer assembly steps, leading to swifter production cycles and lower manufacturing costs. Combined, these benefits mean that despite the current high cost of 3D printers, they are far more competitive versus the whole cost of setting up and running a traditional manufacturing line.

This has already been the case in the production of prototypes for several years, such as for the helicopter blade manufacturer Automated Dynamics, which achieved a 60–70% reduction in the retooling costs for its prototype development.[2] Increasingly, it is also the case in production: while it was building a Suezmax supertanker, Spanish shipbuilder Navantia achieved a 17% cost saving for two ventilation grills by switching from traditional to 3D printing manufacture, despite also changing the material from stainless steel to lighter, stronger carbon fiber-reinforced plastic.[3] In doing so, Navantia reduced the lead time for those items from five weeks to three hours. The UK medical company Crispin Orthotics makes orthoses, which are externally applied devices that prevent or correct disabilities, promote or improve function of the affected area, or help reduce pain. Orthoses may, for instance, assist or resist joint motion, or relieve weight, such as leg braces, insoles, or support casts. Typically, each orthosis is made of multiple parts that need to be assembled. By directly 3D scanning the patient to create the right models, optimizing those for production, then using 3D printers, Crispin Orthotics has reduced the number of individual parts, accelerated manufacturing time, and halved unit costs. The patients benefit from having lightweight, durable, bespoke orthoses in less time than was possible using traditional techniques.

If a broader comparison of traditional techniques versus 3D printing is made, beginning with the development of a prototype, through modifying designs, to producing objects, 3D printing is generally the better option for items that have the following characteristics:

- Final design arrived at after numerous iterative design cycles
- A need for make-to-order or low batch sizes
- Complex design or with high design variability
- Assembled from several subparts
- Made with one or few materials

As discussed earlier in this book, 3D printing allows manufacturers to make better parts, more suited to their circumstances of use. Depending on those circumstances, this includes making items that are stronger, lighter, or more homogeneous. Aerospace companies, for instance, are particularly interested in the potential to reduce weight while retaining strength. The space transport company SpaceX employs 3D printing to produce valve bodies in their engines, having found that the 3D-printed version is physically a better part: stronger and more ductile, less prone to fracture and lower variation in the parameters of its materials. It has also produced them much faster using 3D printing than traditional castings, cutting production from months to less than two days. These benefits are driving NASA's rocket engine parts supply chain, too, which has a stated goal of manufacturing those parts up to 10 times faster and at less than half the original item cost.[4] The space sector in particular is fast adopting 3D printing for other, similar purposes because of potential cost savings. For instance, SpaceX's Falcon 9 rocket costs some US$ 62 million per launch, with a maximum payload of 22,800 kg, resulting in a minimum cost per kilogram of US$ 2,720. The larger Falcon Heavy rocket costs US$ 90 million per launch, with a minimum cost per kilogram of US$ 1,410. This latter rocket is proposed for Mars missions with each kilogram of load costing at least USD$ 5,300. Those numbers make saving every kilogram through topology optimization and 3D printing in new materials a very attractive—and necessary—proposition.

Now consider the flexibility that is required for manufacturing. All too often there is a need to maintain several production lines to cope with changing demand, each line fabricating a different variant, for instance, and the productivity of each dialed up or down according to that demand. 3D printing removes that need, as complexity and variability are more easily handled via the same printer. Of course, different printers have different capabilities, in terms of materials, colors, accuracy, and precision, but where the differences in demanded objects lie in the form of those objects, one printer can cover several versions with no need for the expenses and delays of retooling. Taken to its limit, this opens several opportunities. The company Local Motors, for example, sees the automotive sector using 3D printing's adaptability to return to the earlier days of cars, when coachworkers would design and make bespoke bodies for noted vehicles like Bugattis and Duesenbergs. That, they say, can be done in workshops anywhere in the world, allowing for local tastes to be an input to those designs.[5] Consider the example of Urbee-2, the world's first 3D-printed plastic car unveiled in 2013. Made by mechanical engineer Jim Kor, Urbee-2 is a fuel-efficient, aerodynamic, and lightweight car with the resilience and safety usually found in motorsports vehicles. To make it, Kor 3D-printed 50 parts in plastic, adding a motor, chassis, and safety cage.[6]

Looking at the human resources involved, traditional manufacturing requires teams to make, assemble, and move materials, prototypes, WIP, and finished items at different locations. To tackle that, manufacturing has already seen the advent of robotics, particularly in sectors like automotive manufacture, and has succeeded in cutting errors, improving time, and reducing headcount. As in those industries that have embraced robots, 3D printing requires very few personnel to maintain the machines. Furthermore, as items made traditionally with several parts requiring assembly are replaced by 3D-printed objects that aren't and don't, staff needs will be much reduced.

Because 3D printing requires a smaller footprint, is highly adaptable, and can be deployed reasonably quickly, it offers the possibility of quickly ramping up and operating modern manufacturing in places where deploying traditional manufacturing would be difficult and protracted. Many countries' regulatory and fiscal regimes make importing large manufacturing equipment complicated. Setting up logistics can further delay manufacturing capability, and bigger teams will also be required to install and maintain that equipment. The smaller scale of 3D printing makes it a more attractive option, requiring less space, fewer people, and less paperwork to deploy.

To illustrate this, consider the experience of technological rollout in another sector: telephony. Many countries in Africa, for instance, moved from very basic wired telephony to widespread mobile networks in a very short time, leveraging both their ease of setup and the increased capability that they bring. Places like Kenya then used those networks to develop and deploy world-leading solutions to areas such as electronic peer-to-peer payments, with examples such as M-Pesa (the "M" standing for "mobile" and "Pesa" being Swahili for "money"). This is a mobile-phone-based money transfer, financing, and micro-financing service launched by Vodafone for Safaricom and Vodacom, the largest mobile network operators in Kenya and Tanzania, in 2007 and since expanded to other developing economies in Africa, Asia, and Europe. This service has brought advanced maturity in financial transactions to those countries in a way that would have taken decades otherwise.

Likewise, 3D printing can quickly bring high-quality, responsive manufacturing to developing economies, such as the experience of ReFabDar in Tanzania discussed in Chapter 2. Organizations in the humanitarian and defense sectors have already grasped this and are developing deployable solutions for their fields. Both industries and developing economies' wider societies also stand to gain from fast-developing advanced manufacture.

The rising need for sustainable manufacture, whether through reducing wastage, improving recycling, or better proximity to a circular economy, means that there is also a significant body of research into how to minimize the environmental footprint of 3D-printed parts. Unused materials in powder-based techniques, such as selective laser sintering (SLS) and selective laser melting (SLM), can be recovered and reused. Several companies are researching materials for 3D printing that can be recycled, a key element of the closed-loop model. However, much progress is needed to achieve that goal at industrial scales. Still, it is already clear is that 3D printing is by far the most sustainable manufacturing solution.

Looking at the SCOR metrics set, those attributed to manufacturing that will be most affected by 3D printing are as follows:

- *Production Cost* is reduced due to lower production labor; property, plant, and equipment costs; and a simplification of the production lines.
- *Make Cycle Time* and *Retooling Time* are both reduced.
- *Wastage Rate*, the percentage of materials that are wasted during manufacture, is reduced in many of the 3D printing methodologies, as unused materials can be returned to the process. This metric is particularly important in those sectors where the costs of raw materials are especially high, such as those using precious and high-grade metals.
- *Carbon Footprint*, stemming from lower energy consumption throughout a 3D-printed object's production cycle. However, a fuller, end-to-end analysis of energy consumption and consequential emissions may indicate that those are merely shifted elsewhere in the value chain, such as in the production of raw materials.

An Aid to Postponement

It has long been the case that a way to manage variations in item specifications across a range of customers is to employ postponement in the supply chain. These variations can stem from differing personal tastes, languages, or national or company standards, for example. Rather than have dedicated production lines for each variation, a postponement strategy calls for a common platform to act as the base, and the final components, modules, or alterations required to make the end product are added later or closer to the customer. Companies such as the clothing firm Zara and the IT provider Dell have employed supply chain postponement to such a degree that they have set the benchmark for responsiveness in their respective sectors. Several parts of the supply chain can use postponement, from manufacturing through assembly, to packaging and labeling.

By employing 3D printing in their postponement strategies, supply chains can combine the benefits of both. While 3D printing offers the possibility of producing lots of one, the effort to customize each design every time is onerous and expensive. Using a base design from which to make smaller changes, supply chains can offer the variety they seek without the added costs. For example, firms can provide a range of base designs to consumers, who can then further personalize them. As part of the events for its 125th anniversary, consumer electronics firm Philips partnered with the German company Twikit to offer 125 of its customers the opportunity to personalize their face shavers. Using one of the latest Philips models as a base, each person could select the color, texture, and form of the external grip from a palette of options on Twikit's platform, which were then produced at Philips's 3D printing facilities. This allowed Philips to provide a distinctive customer experience with minimal impact on their production costs and lead times.

Naturally, this model can be used with logistics providers and local distribution centers, which can be given inventories of near-complete items to hold and then finish them with 3D printers to produce the end-customized parts.

3D PRINTING AND "DELIVER"

After reducing the cost and time to manufacture, the next wave of industrial benefits of 3D printing will be in the area of logistics: storage and movement of materials, parts, and products throughout the value chain. For a supply chain to be effective, it must ensure that the right things are at the right place at the right time, and that can quickly become a very complex demand. If it fails, then the customer doesn't get what they want, and the business spends more to recover from that failure. If a spare part is not available when and where it is needed, machinery is not repaired, production is held up and again, the customer is again disappointed, and the business spends more in recovery. Traditionally, companies have developed expensive distribution networks, with costly transportation and warehouses filled with materials, partly completed products, finished items, and spare parts, spending heavily on those while locking in working capital that could be used more efficiently elsewhere. Many firms go further and invest in optimization to make those operations as efficient, effective, and economical as they can be. 3D printing offers another way. The transformation that 3D printing enables manifests itself in three areas:

- Demand for warehousing
- Need for transportation
- Exposure to supply chain risk

The savings are significant: a 2014 study by the consultancy PwC estimated that adopting 3D printing to replace maintenance, repair, and operations (MRO) spares in the aerospace sector alone would save US\$ 3.4 billion annually, based on half that inventory shifting to 3D printing.[7] Even if that assumption were to be a reduction of only 20%, that still represents over US\$ 1 billion in that sector alone.

Inventory is a complicated and complex matter that can easily make the difference between a business's success or failure. Too much inventory and the business locks up working capital and becomes exposed to higher storage and insurance costs. If those items are held for too long without being used, they will be depreciated and eventually written off, and throughout they are at risk of becoming obsolete or surplus to requirements, thereby requiring disposal and incurring further costs. Too little inventory and the supply chain risks failing to meet its required service levels and, in turn, customers' expectations, damaging reputations and future revenues. In response, firms expend considerable effort in refining mathematical algorithms that seek to optimize inventory and balance the service levels required, material lead times, and myriad other factors. Others have sought to increase the flexibility of their manufacturing and supply chain processes, seeking to keep those agile in the face of changing demands. 3D printing offers an attractive alternative—making items on demand.

Rather than storing a long tail of parts, WIP, and finished goods, many supply chains—such as Deutsche Bahn's that was described in the previous chapter—are already considering 3D printing to reduce the need to hold those by digitizing them and producing them on demand, effectively switching their inventory policy to a make-to-order model. The use of on-demand 3D printing enables the virtual warehousing model: rather than holding many thousands of items, only a few are needed, with the ability to quickly replenish stock within a matter of a few hours or days. That reduces the need for storage space, lowering the cost of stockholding, and it frees up working capital; a 20,000-square-foot warehouse with over US$ 20 million in inventory could be replaced with a bank of 3D printers costing US$ 2 million in a 200-square-foot on-demand facility. It was just these benefits that initially drove British audio equipment manufacturer Bowers & Wilkins to decide in 2013 that it was going to use 3D printing to pare its inventory levels while preserving, if not improving, the availability of parts to its service department by producing them on demand.[8] Traditionally that policy was fraught with the risk of manufacturing delays threatening timelines, and this could still be the case with 3D printing. However, if viewed from an end-to-end perspective, from order to fulfillment, the technique could be advantageous for scheduling if elements such as logistics are included.

Spare parts are a particularly attractive first area to tackle with 3D printing, as they tend to be slow moving, particularly those required for MRO. The design specifications used to make them in the first place are frequently not available—all too often, that data was never requested when they were ordered, or the original supplier stopped making those parts or is no longer around. Larger firms making or using complex products, such as jet engines, may have tens of thousands of these parts on hand, with the financial exposures that that involves. As those parts age, many will degrade and weaken, becoming unfit for use and increasing the cost of obsolescence. Several firms, from Airbus to Shell, have noted these characteristics, and are now implementing projects to produce those parts on demand to lower costs and improve service. Porsche Classic began to 3D-print spares for some of its classic cars in 2017, such as the clutch-release lever for the 959 and a crank arm for the 964.[9] These parts were no longer available to reproduce or in stock anywhere, and restarting production lines would have been costly for both the company and the enthusiasts needing the parts. Instead, the spares are made on demand to the same standards as the originals, mitigating the need to stock quantities of parts that may never be called for while giving the customers access to them virtually on demand. Interestingly, Porsche found that the 3D-printed parts exceeded the required build standards, echoing the experience of other manufacturers.

Clearly, the viability of 3D printing spare parts will depend on having a good business case that demonstrates the value in that approach. Factors such as what the spares are made of, the availability of suitable 3D printers, and the levels of investment needed compared to the benefits that would be accrued all contribute to that, and they may be found wanting.

Along with a reduction in warehousing, the need to physically move parts is also significantly reduced. Given that 3D printing permits software files to be transmitted or emailed to the printer, versus moving physical inventory, manufacturing or delivery of finished products can originate far closer to end users. This is achieved by placing or using printers closer to customers, either within a company's own printing facilities or via outsourcing to a nearby 3D print hub or bureau. This benefit has been ably demonstrated by Made in Space's example with the International Space Station, seen in the Introduction, and it radically changes how Delivery Lead Times (DLT) will be perceived and what will be expected of suppliers. No longer will parts need to be stored in vast storage spaces or shipped across and between countries, with 3D printing reducing DLT from weeks to hours.

This shift in logistics will significantly affect those companies that provide warehousing and logistical solutions. It constitutes a huge threat to third-party logistics (3PL) firms' very existence, and the sector is now looking at how to adapt, with many opting to act as 3D printing manufacturing centers. According to a survey conducted by the business intelligence firm EFT, the proportion of 3PL firms already offering 3D printing expertise or services, or considering doing so, rose from 24% to 41% in the 2014–2015 period alone, and that trend continues to gain pace.[10] West Monroe Partners consultants Aaron Bresinger and Jeff Arnold said of 3PLs:

> "E-commerce shipping volume may suffer as sellers recognize the capability to transfer their product designs electronically for 3D printing at a location near the consumer. To mitigate the risk of losing shipping business to the electronic transfer of 3D printing blueprints, industry leaders need to incorporate 3D printing into their strategic thinking."[11]

The replacement of obsolete parts is also driving the adoption of 3D printing in several capital-intensive sectors, such as oil and gas, mining, transport, and automotive, industries where parts can have prolonged lifetimes during which their manufacture may have ceased, or the original supplier might have gone out of business. The consortium Mobility Goes Additive, seen in the

The Spare Parts Case

Imagine a company that has a requirement for one steel strut. Say the market cost of the mass-produced item is US$ 8.50, and that the minimum order quantity is 10 struts, for a minimum spend of US$ 85, and that it has a 10-year lifetime. The supplier's lead time may be weeks. Once the single strut has been installed in the equipment where it was needed, the other nine are stored on the shelf and may well never be needed again. During that time, they are taking up space in a warehouse, clocking up costs in storage, insurance, and personnel. While there, those nine struts will depreciate and ultimately incur some disposal cost. It could well be that the warehouse is not located close to where the struts are used, and there will be logistics costs of getting them to where they're needed. These costs soon add up, and the original US$ 8.50 cost for a strut can balloon to over US$ 100.

This situation is typical for spare parts (Figure 6.2). Demand for them is unpredictable and complex, instead of the smoother profiles that "normal" items enjoy. They move more slowly but when they're needed, many will be essential, and the service levels for those will be above 99%. They typically have longer lead time, particularly when remote operations are involved. Often, they can only be sourced from a few suppliers, potentially only one, which can incur higher taxes and tariffs if they are foreign. The result is that more parts are held in stock, resulting in a long inventory tail and further increasing costs.

FIGURE 6.2 The dynamics of spare parts.

(continued)

(continued)

3D printing offers a better alternative, giving companies a way to retain high service levels without the need to stock hundreds, if not thousands, of parts that may never be required. Returning to the strut example, even if the cost to design and 3D print the strut is three times the original cost, say US$ 25.50, this is still far cheaper than the alternative. Moreover, it can also be produced on a 3D printer closer to the point of need, thereby reducing distribution costs and lead times. If the situation with the strut is repeated for hundreds or thousands of other parts, the mathematics of 3D printing make it a clear choice for spare parts supply chains.

With spare parts being an obvious choice for 3D printing, new firms have started to focus specifically on this area. The company Kazzata was founded by Noam Eshkol and Avichai bar Lev in 2014, who leveraged their experience of managing repositories of stock photographs to create a company that acts as a cloud-based marketplace for spare parts. Kazzata provides manufacturers with a platform that stores and manages digital design files of spares, selling, making, and shipping them without the need to do so themselves. This is done using IP protection systems and industrial 3D printers to produce the spares on demand.

Deutsche Bahn case example in Chapter 5, came together to actively pursue opportunities to reduce obsolescence in industrial spare parts in addition to other more traditional avenues. Instead of paying a significant premium for an old supplier to retool to make a replacement, or to a new supplier to produce that part from scratch, the firm can resort to a design company to scan the obsolete part and generate a data file (which naturally would require understanding and re-engineering the internals of the required part) and pass this to a suitable 3D printer for fabrication.

Clearly, therefore, the logistics metrics that will be most affected are those related to cost, time, and customer satisfaction:

- *Inventory Costs, Transportation Costs* and, from those, *Fulfillment Costs* are reduced as the need to hold and move stock is lowered if not eliminated.
- *Delivery Performance to Customer Commit Date, Delivery Lead Times (DLT)*, and from that, *Order Fulfillment Cycle Time (OFCT)* are reduced by placing manufacture closer to the end consumer of a product.

- *Customer Satisfaction* is improved, at least initially, as customers who are used to long lead times are now catered to much faster.

These metrics will only improve markedly in the first few periods that 3D printing is used, owing to diminishing returns, and as warehousing is space is optimized, as lead times return to the lowest practicable level and as customers become used to the new dynamics.

3D PRINTING AND "RETURN"

The return part of the supply chain is increasingly important, driven by both economic and regulatory needs. Today, it encompasses not just the return of goods sold (the so-called reverse supply chain), but also aftersales and customer support, warranty-covered repairs, and the processes associated with any products returned to a company's supplier. That includes issues such as through-life upgrades and modifications, replacement, and obsolescence, areas that can be quite costly to manage if done badly and profitable if done well. Many involved in these areas of the supply chain are already actively investigating, if not actually using, 3D printing to offer better levels of service and lower cost.

For example, consider those industry sectors that face issues of equipment and parts that have long lifecycles, such as naval ships and submarines, which are typically used for several decades. The same can be said of oil and gas platforms and much of the major equipment in the mining and transport sector. As mentioned previously, those long lifecycles require optimized inventories of spare parts to keep the equipment operational. Therefore, those sectors have vibrant and expensive aftersales sectors, offering warranties, servicing, and parts. Pricing those is often complex, requiring the consideration of safety stocks, costs of storage, probabilities of parts failure, and so on, all to arrive at the right figure to charge customers. If parts are 3D printed, however, from the same digital files that made them in the first place, the need for that aftersales machine diminishes. This has wider implications: the Waste and Resources Action Programme (WRAP) has stated that, in the UK, about a quarter of electric equipment ending up in household recycling centers can be reused with small repairs.[12] If the right spares can be made available where and when needed, then products last longer and become cheaper to repair, reducing landfill and disposal costs. Similarly, the need to reduce obsolescence through expensive part upgrades is also simplified, as tweaking the digital design and producing the "new and improved" part is faster and ultimately cheaper.

That reduction in the burden of obsolescence and the need for spare parts both stem from the ability of 3D printing to shrink component inventories by making, in one unit, parts previously composed of several items. Imagine a product that comprises 10 individual parts, each of which may require replacement. As discussed in Chapter 2, if the entire product is made as a single unit, essentially moving the level of lowest replaceable part, then there is no need for an extensive inventory of separate parts. This advantage must be balanced against the consequence of moving the point of lowest replaceable part—it might become necessary to replace the whole item, versus one or more parts, which makes through-life support more expensive and wasteful.

As a consequence of these benefits, companies see a reduction in levels of required inventory, therefore reducing the working capital that is tied up in that inventory. With that reduced inventory comes less wastage, as a long tail of items purchased up front in case of need won't have to be scrapped when real consumption is below forecasted demand.

As new materials become increasingly recyclable, the potential for a 3D-printing-enabled closed-loop supply chain develops. Many items are heat-processed plastics and, as Hans-Georg Kaltenbrunner, Vice-President, Manufacturing Strategy EMEA at JDA Software, notes:

> "[it is] possible to create a reverse supply chain approach. Customers can recycle used, damaged or unwanted goods by taking them back to their local 3D printshops so that they can be melted back down and made into something new and useable once more."[13]

However, the practicalities of that may be impinged as more complex and multi-material objects are produced. Recycling the sorts of cartons made by firms such as TetraPak, Elopak, and SIG Combibloc, for instance, which typically comprise layers of paper, polyethylene, and aluminum, is a difficult and expensive task. Recycling multi-material 3D-printed objects may similarly be uneconomic.

In the Return element of the supply chain, the most affected metrics will therefore be:

- *Cost of Warranties* and the *Cost of Obsolescence*, as there is a reduced need to hold spare parts to cater for warranties and to retain underused production facilities to manufacture components. Moreover, as designs can be modified prior to their manufacture more easily, spares and equipment are less likely to become obsolete as their designs evolve over time.

■ *Disposition Costs*, driven by the lower need to hold large numbers of spare parts that are not used in the lifetime of the equipment they support.

3D PRINTING AND "SOURCE"

The impact of 3D printing on sourcing lies in three key areas:

■ What is procured
■ How it is procured
■ How much it is procured for

What Is Procured

It isn't surprising that 3D printing requires the procurement of different things. To start with, raw materials for 3D printing are different from those for traditional manufacture. Rather than buying solid metals, for instance, metal powders are needed. Where items are to be made in thermoplastic, that plastic usually needs to be available as a fiber with the right properties for 3D printing. This alone is a significant change to procurement. The providers of 3D printers and printing services will be new to the procurement department. Of course, where the design of items simplifies them, reducing the number of component parts, this also reduces the procurement burden to acquire those components. These sorts of changes are easily adopted by procurement departments.

For larger, more complex items and equipment, procurement needs to consider spare parts, their obsolescence, and through-life upgrades, all of which must be negotiated and priced into the deal. This can be up front or a long time after the main item was bought, particularly in capital-intensive sectors like defense, oil and gas, transportation, and mining. A warship typically has a lifespan of 15 years, an offshore platform 30. Buying the lifetime supply of spares requires procurement teams to model several complex factors of consumption, operation, reliability, storage, and obsolescence to arrive at a number that will inevitably be wrong. If spares are left to be bought as required, separately from the main equipment, their costs can soon explode when retooling or recommissioning suppliers is added to the cost of the actual parts. 3D printing reduces this exposure, simplifying procurement tasks and reducing their costs through on-demand production by increasing the inventory days of supply without incurring the costs of holding that stock.

Underlining this is the increasingly important role that procurement will play in understanding which items are purchased from third parties as an input to analyzing which would be better to 3D print. Factors such as lead times, necessity for assembly, transportation costs, and so on will all influence the evaluation process, more closely involving procurement teams in the operational decisions of the business than before. The late David Noble, formerly the group CEO of the Chartered Institute of Procurement and Sourcing, told an audience in London in 2016 that the growth in 3D printing's role in supply chains puts greater emphasis on procurement, saying:

> "It means that a shift from manufacturing to materials acquisition becomes a core differentiator."[14]

Often, though, the data needed to conduct the required analyses is not easily available, requiring the scouring of contracts and IT systems to build it, as well as filling in gaps through the judicious use of the procurement team's experience and insight. In many companies, however, this need for a collaborative approach will be hampered by functional silos that will need to be dealt with accordingly. Independent of whether to engage with 3D printing, and in light of the emerging trend for data analytics and digital transformation, the earlier a procurement department begins to build these datasets and gain clarity of what it is acquiring, the more prepared it will be when the questions are asked.

How It Is Procured

The biggest impact of 3D printing on actual sourcing comes from another change: the shift to procuring data rather than physical items. As supply chains become more digital and companies trade based on digital designs of items that are then 3D printed, the size, shape, and processes of procurement will change. The approach to procuring physical things is different from acquiring data-based designs. For example, it requires new ways of carrying out supplier and product assurance, and new questions to be asked of their suppliers. As when procuring software, this makes evaluating suppliers and their designs more elaborate, involving questions of configuration control, information security, limitations, and requirements of use that go far beyond the norm when buying physical items. Licenses for using the data will specify who can use the data, in what circumstances, on what equipment, and for how many iterations. The obligations of intellectual property and liability will be defined in terms that are broader and more detailed than is usually the case with hardware. Procurement cycles will tend to be longer and procurement teams will need new skills. The capabilities that 3D printing presents also change some fundamental principles of procurement. The ability to produce "lots of one," as well as multiple lots on the same machine with few or no changes to the tooling, means that there is little need for minimum or economic order quantities—the same data file can be used to make one item or many. It is the license that regulates how many of an item can be produced from the same data file. The changing nature of procurement raises the question of how the licenses will be enforced. Lessons can be learned from the music industry, which went from selling vinyl records, cassettes, and CDs to selling audio data files in the last two decades. During that uncontrolled transition, the industry went through considerable difficulty, particularly with respect to the payment of royalties, as unauthorized copying of music was rife. Manufacturing has an opportunity to tread over a used path and ensure that it doesn't fall over the same hurdles with 3D printing.

How Much It Is Procured for

As well as being simpler to generate, the pricing structure of purchased 3D-printed items is more predictable. This arises from several factors:

- The simplification relative to the pricing models for equipment, spares, and support as supply chains become more digital;
- The lower complexity in the supplier portfolio, which results from having fewer separate components in a final product; and
- The ease of updating and modifying data files rather than reissuing physical items.

Although sourcing software can generally be a more complex and involved process than sourcing physical things, that stems from the need for long evaluation phases. With 3D printing, what is acquired is a design file for a part, one that usually will have developed with the end user. With the development of standards and regulations for 3D printing (see Chapter 8), the effort and time required to build those design files will be considerably reduced. Once a design is approved for purchase, further procurement will be much faster.

Another procurement cost benefit is the reduction, if not elimination, of cross-border tariffs. Several countries across the globe impose import duties on parts, WIP, and finished goods. With 3D printing, the subject of the transaction is software—a data file rather than objects—so until the system of tariffs is changed and a means to regulate it for software developed, these files can be passed across borders without incurring those import duties. This eliminates the need for complex Incoterms such as FOB (Free on Board), CPT (Carriage Paid to), and the like. For example, Brazil today imposes a levy of 18% on the importation of several stainless-steel parts. If users of these parts, like Ford, decide to employ local 3D printing facilities and transmit the data files from the USA or Europe for parts to be manufactured in Brazil, their cost bases will decline—with an accompanying loss of revenue for the Brazilian government. However, more advanced economies may be catching up with this. The European Union adjusted its legislation on value added tax[15] to cover telecommunications, broadcasting, and e-services that are sold directly to customers; it is realistic to consider this being extended to include transactions of digital design files in the near to medium term.

From a sourcing perspective, then, these will be the most affected metrics:

- *Sourcing Cost*, as resource requirements are more likely to decline, since it is easier to carry out the procurement activities of software than physical items, and the additional costs on top of a part's base cost due to tariffs, life spares, and so on also decline.
- *Inventory Days of Supply* are increased, but without the added stockholding costs.

3D PRINTING AND "PLAN"

In most industry sectors, it is recognized that the right supply chain is a strategic differentiator, and getting all the parts of the supply chain working correctly, within and between organizations, is critical to success. That means planning the supply chain is now of paramount importance. Such planning is increasingly affected by customer requirements, and those can and do change rapidly, particularly in consumer goods sectors, but also in capital-intensive industries. Once those needs are understood, having a supply chain that is agile, able to react quickly to changes, will ensure those dynamic requirements can be met. Moreover, if the supply chain is to react in the right way and at the right time, it needs to operate in an "always on" environment, not only bringing visibility of what is taking place along the whole value chain, but also reacting to changes in real time, or at least as close to that as is practicable. Supply chains working with traditional manufacturing simply cannot meet all of those needs in the emerging fast-changing market without considerable investment in capacity and data analytics, but those using 3D printing can.

Planning a supply chain involves understanding three things:

- What the strategic goals of that supply chain are
- What the customer wants to see from it
- How the business is going to meet that need

Typically, the output from those factors informs decisions of where to place the elements for the entire supply chain, from manufacturing to warehousing facilities: analyzing which provider to use for logistics; what the necessary inventory levels of raw materials, WIP, and spare parts are throughout the supply chain; and what the interrelations among manufacturers, suppliers, customers, and the company itself should be. At the heart is answering how to best service the customer while balancing cost, quality, timeliness, and complexity. 3D printing affects all these parts, such as enabling a reduction in inventory levels by moving products out of physical storage into a digital state, and changing the philosophy of a supply chain, from make-to-stock to make-to-order. This reduces the warehousing footprint needed within the modern supply chain—no longer do we need warehouses filled with long tails of products and parts that have very slow turnovers. 3D printing will also enable a reduction in the number of retail and support centers; imagine customers no longer needing to go to a shop to buy a product or a part, but rather moving toward an Amazon-like operating model where they select products and parts online, customized, ordered, manufactured to their specific requirements and delivered to their door as and when required (within the limitations of the 3D printing solutions available). As we move into local manufacture, we can go to a local fabrication center to buy that final product instead of a shop.

In our ever more competitive world, supply chains are increasingly aiming to achieve perfect orders, getting the right thing to the right place and at the right time. This trend is particularly prevalent in the retail and service industries, where customer dissatisfaction is communicated quickly, particularly in today's environment of peer reviews. That trend is also growing in industrial sectors as customers are less tolerant of mistakes than ever, usually because it eventually costs them. Getting orders right also reduces the incidence of returns, a concern usually forgotten in many supply chains that is wasteful and often expensive to deal with. The advantages of 3D printing, with the ability to ensure the right thing is made quickly and close to, or even at, the location where it is needed and to do so "on demand" make it a valuable and desirable channel for manufacturing.

Supply chain strategies are shaped by a company's business strategy and naturally vary accordingly. Amazon has a stated aim of getting its products to the customer with the "lowest prices" and the "utmost convenience," which has led it to include technology innovations in its logistics. In the USA, it recently filed to use drones to deliver products in under an hour in urban areas, going further in the employment of predictive logistics and sending out goods before they are demanded, all of which is dependent on excellent demand algorithms. In the early 2000s, when Nokia opted to become the fastest producer of mobile phone handset models, it rearranged its market research, suppliers, manufacturing approach, and logistics framework to accommodate that aim. Other companies have chosen to focus on service levels, ensuring that critical products are available at near-zero notice, and requiring the establishment of distribution centers near their customers. The advent of 3D printing enables the possibility of new supply chain models, as described in more depth in Chapter 7.

One part of supply chain planning that 3D printing has already affected is market research (Figure 6.3). Traditionally, a firm will employ researchers to develop a requirement based on customer feedback, after which a prototype will be made, then given to customers to test and comment on; that feedback is incorporated into a new iteration of the requirement and the product. In this scenario, volumes of products are low, so the costs per item can be quite elevated. 3D printing reduced the time and cost involved.

Customers can participate in the digital design stage far more readily by receiving a prototype and commenting on it, and that feedback will be incorporated in the next design and prototype in a fraction of the time, and therefore the cost, of the traditional approach. This has parallels with the Agile design approach that software engineering has used for the last 15 years. As opposed to the so-called traditional, or waterfall, approach, Agile aims to get a design to the requirement though frequent and constant feedback from the end user, moving the design on a little and testing that with the user, then using that feedback in the next iteration, and so on, until the user is happy with the result. Retail companies are already using 3D printing for such approaches in market research to pilot new ideas with the buying public. Unilever regularly uses it to test new packaging forms and textures more economically and effectively. With 3D printing's capability of producing short runs, those pilots can quickly be produced in batches for more rapid, direct, and wider customer feedback. When GE's Oil and Gas division began to use 3D printing to help with their prototyping, they reduced the cycle time from design to prototype from 12 weeks to 12 hours.[16] With these faster research cycles, manufacturers can respond to

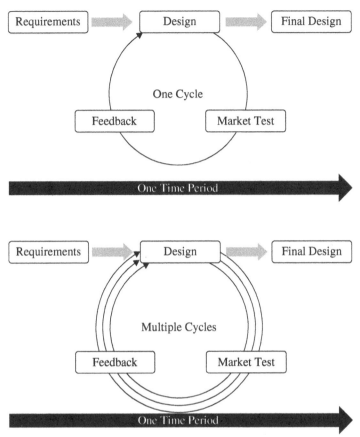

FIGURE 6.3 Comparison of design cycles using traditional (top diagram) versus 3D printing (bottom diagram) approaches.

changing demands more quickly and accelerate product lifecycles, giving them a competitive edge in what are frequently crowded markets where even the smallest advantage is the difference between being a winner and an also-ran. The biggest beneficiaries of this in the shorter term are new market entrants: if a new player wants to move into a sector, no longer are they going to have the burden of a high cost of entry, and a lengthy and expensive period to get the product "just right." Whereas before they would have to secure a prototype manufacturer, and pay for tooling them up with equipment that may have been bespoke, they can now move from design to first production in a matter of days as opposed to weeks, and for hundreds of dollars as opposed to several tens of thousands.

Instant Supply Chains

One set of supply chains that are already being transformed by 3D printing are those concerned with the humanitarian and emergency sectors. International aid organizations like Médecins Sans Frontières (MSF), the Red Cross, and the Red Crescent have used 3D printers to provide prosthetics and other medical equipment, pharmaceuticals, spare parts, and even shelters in areas where logistical difficulty would prevent those reaching people in need quickly enough. Many of these organizations don't have a permanent supply chain network, instead relying on quickly establishing these when called. One example of this is Field Ready, which works with humanitarian organizations to provide engineering-based solutions to supply chain issues, and uses 3D printing to provide "supplies in the field."

When the US Virgin Islands were struck by two Category 5 hurricanes in late 2017, the island's electrical systems were destroyed, seriously hindering efforts to get aid to those who needed it as well as curtailing everyday life. Because efforts to restore the islands' power generation were lagging, the team from Field Ready collected as many solar panels as they could find to provide electricity more swiftly. Many of the panels were found to be functional, but they needed to be charged, and with the wider power network not available, an alternative had to be found, using high-capacity batteries instead to give them a boost. To do this, the panels needed a bespoke part to connect them to those batteries, and the lead time to get that was too long. The Field Ready team created a design for the connector using computer aided design (CAD) software and produced it on a 3D printer, itself using industrial batteries.[17]

Before this, in one of the first examples of 3D printers being used in postdisaster recovery, Field Ready produced a fitting to repair the water pipes in a displaced peoples camp in Nepal in the aftermath of the magnitude 7.8 earthquake there in 2015, which killed 9,000 people and injured over 21,000. The plastic part was produced virtually overnight using a 3D printer powered from the battery of the Field Ready team's Land Rover. Once it was installed, the 200 families in the camp had access to water.[18]

To NGOs, the advantages of 3D printing are plain to see. They enable the fabrication of items at or very close to the locations where they are most needed, using industrial or car batteries to power them. This advantage is amplified with the use of drones to distribute them. 3D printers have the flexibility to make a wide variety of parts customized to a specific person (in the case of a medical part or prosthetic). The effects of disasters and emergencies mean that items needed for repairs either can be one of a huge range, or bespoke, depending on what equipment was broken or damaged. Therefore, rescue, recovery, and support teams must carry or convey a large amount of gear to cover all options. Using 3D printing to make many of those items lets teams reduce this load. At the same time, the usual constraints of 3D printing, such as the need to machine them to a high level of finish, are superseded by the need for a good-enough, possibly temporary part. This all translates into lives saved and improved and costs reduced. For instance, a 10-cent umbilical clamp needed in Haiti, which was hit by catastrophic and consecutive floods and earthquakes, costs US$ 1.00, of which 90% represents transport and storage. In 2015, Andrew Lamb from Field Ready calculated that even if the savings in these logistics costs were in the order of 40 to 50%, that would cut between US$ 5 billion and 7 billion per annum from the costs of humanitarian aid.

Given 3D printing's ability to make things on demand, one area of supply chain management that will be eased is sales and operations planning (S&OP). Integrating management processes into all of an organization's functions, S&OP uses an up-to-date forecast of demand to tune the sales plan and, from that, the production, inventory, and other supply chain plans. The goal is to balance all of those, resulting in an efficient, effective, and economic supply chain. However, S&OP is often plagued with inefficiency because of the difficulty of properly forecasting demand. The result is a misalignment between what customers, internal and external, want and how the rest of the supply chain is set up to meet that demand. One obvious solution is to have highly accurate demand forecasts, and organizations can expend considerable time and resources to achieve that. A more practical approach is to improve the flexibility and responsiveness of the supply chain; it is here that 3D printing can help. By responding more quickly to changes in required volumes of different configurations and designs, and rapidly reconfiguring what is produced, 3D-printing-based production mitigates inaccuracies in forecasting, enabling that goal of a good supply chain. Other contributions to an effective S&OP derive from 3D printing's effect on the other areas of the supply chain described in this chapter.

The success of the plan element of the supply chain is typically conveyed in four key metrics, which are all improved by 3D printing:

- *Total Cost to Serve (TCTS)*. As 3D printing reduces the costs of the individual elements, from market research and full manufacturing to the reduced need for warehousing and distribution, it significantly reduces TCTS.

- *Perfect Order Fulfillment (POF) and percentage of orders delivered in full (PODIF).* 3D printing allows the customer to see exactly what they will receive and, provided that the elements of the supply chain are suitably synchronized, errors will be minimized, improving PODIF and, from that, POF.
- *Order Fulfillment Cycle Time (OFCT).* 3D printing accelerates the individual stages of a supply chain, from market research to delivery. When one considers the full time for manufacture, from tooling up to production, OFCT is significantly reduced. Of course, that benefit abates as volumes increase, and traditional manufacture will be the dominant approach for several decades until 3D printing enters mass production. However, even in those situations, using 3D printing to accelerate mold production time brings advantages overall.
- *Supply Chain Flexibility and Adaptability.* As 3D printing can cope with changes in form significantly faster than traditional manufacturing, and as manufacturing can be shifted more easily to available capacity elsewhere, both the flexibility and adaptability metrics are improved.

3D PRINTING AND "ENABLE"

The enable area of the supply chain refers to the activities associated with the management of the supply chain, such as the management of:

- Business rules
- Performance
- Data
- Resource
- Facilities
- Contracts
- Supply chain network
- Regulatory compliance
- Risk

3D printing requires a different approach to shaping and managing the supply chain, which affects all these directly or indirectly. However, the scale of the differences is unlikely to be significant: after all, all the preceding areas apply equally and have value whether using 3D printing or traditional manufacture. Earlier parts of this book have touched on several of these changes, such the need for different approaches to contracts and procurement, data management, and the supply chain footprint. As with all situations where companies introduce new techniques, like Lean, Six Sigma and just-in-time, the management of these activities needs to adapt to the new reality.

One activity area is worth highlighting in more detail: that of supply chain risk. The 2011 disaster in Fukushima, Japan, caused by the cataclysmic combination of the Tōhoku earthquake and tsunami, and the consequent Fukushima Daiichi nuclear accident, significantly reduced the country's electricity supply. The tragedy halted 90% of production capacity in the area, including the silicon wafer and automotive parts supply chains there, which comprised 22% of the world's 300 mm wafer production and 60% of its critical automotive parts industry,[19] as well as a significant percentage of other products such as flash cards.[20] Similarly, a fire at a Royal Philips Electronics plant in Albuquerque, New Mexico, on March 17, 2000 threatened the supply of chips to two of the company's major clients. Between them, Nokia and Ericsson accounted for 40% of the plant's demand. Without this plant, production of several million handsets at both would come to a shuddering halt. It took industry-changing efforts to mitigate the impacts, with the bigger lessons being the need for supply chains to quickly adapt to new circumstances.[21]

These are the sorts of risks that many supply chains face today, with manufacturing restricted to single or a few locations. If those are impacted by any one of several threats, natural or human in origin, the time needed to recover can slow the entire supply chain. The identification of supply chain risks needs to go deeper. The lack of a simple item costing a few dollars could halt operations for hours or days, with revenue risk measuring in the hundreds of thousands or more per day, particularly in sectors like oil and gas or mining. Fortunately, today there is increased recognition that supply chain risk management is essential to successful and sustainable operations, and more companies have processes and structures in place that seek out and reduce risks in each part of the chain. 3D printing mitigates many of those risks. It allows for items to be made to order, closer to where they are needed, reducing lead times and providing a way to mitigate the threat of an unexpected spike in demand. As production can be quickly redirected away from threat-rich locations, and distributed more widely if needed, 3D printing makes supply chains more resilient.

In the enable set of processes, therefore, the most affected metric is:

- *Supply chain risk*, which is lowered, together with a potential reduction in contingency budgets, as supply chain elements, from sourcing through manufacturing to logistics, can be more easily distributed across geographies to minimize the potential for single points of failure in the network.

THE SUPPLY CHAIN CASE FOR 3D PRINTING

Table 6.1 presents a summary of the how 3D printing impacts the various supply chain metrics. Given the benefits that 3D printing brings, it is clear that businesses cannot afford to ignore it, nor think it is something that can be postponed for a decade. Most company strategies look at a planning horizon of 5 to

TABLE 6.1 Summary of supply chain metrics affected by 3D printing.

Supply chain process	Metric	Impact of 3D printing on metric
Plan	Total cost to serve	↓
	Perfect order fulfillment	↑
	Percentage of orders delivered in full	↑
	Order fulfillment cycle time	↓
	Supply chain flexibility and adaptability	↑
Source	Sourcing cost	↓
	Inventory days of supply	↑
Make	Production cost	↓
	Make cycle time	↓
	Retooling time	↓
	Wastage rate	↓
	Carbon footprint	↓
Deliver	Inventory costs	↓
	Transportation costs	↓
	Fulfillment costs	↓
	Delivery performance to customer commit date	↑
	Delivery lead times	↓
	Order fulfillment cycle time	↓
	Customer satisfaction	↑
Return	Cost of warranties	↓
	Cost of obsolescence	↓
	Disposition costs	↓

10 years, and in that time 3D printing will be a more prevalent component of supply chains. Indeed, the companies that will succeed in the medium term are those that realize the benefits described in this chapter: increasing revenues and reducing costs.

These factors have been echoed by companies pressing ahead with 3D printing. A survey by the firm Sculpteo in 2017 found that 70% of companies they interviewed had increased their investments in 3D printing, up from 49% the previous year, and more competitors were also using 3D printing (74% vs. 59% the previous year). In the companies that were using the technologies, 43% were using it for production, up from 22% the previous year.[22]

The impacts of 3D printing on individual elements in the supply chain are wide and significant, and lead to operational improvement, even excellence. However, to make significant step changes in performance, operational innovation is called for, doing things in a completely new way. 3D printing enables that innovation by facilitating the implementation of new supply chain models. This gives not just companies, but entire sectors, a means to make that leap and do things differently. Chapter 7 describes what those emerging supply chain models are.

Notes

1. Shoshanah Cohen and Joseph Roussel, *Strategic Supply Chain Management: The Five Core Disciplines for Top Performance* (New York: McGraw-Hill Education, 2013).
2. Carrie Wyman, "Helicopter Blade Prototype Tooling Costs Reduced by 70% with Stratasys 3D Printed Soluble Cores," Stratasys blog, June 15, 2015, http://blog.stratasys.com/2015/06/15/helicopter-blade-prototype-3d-printing.
3. Beau Jackson, "3D Printing Saves 17% on Navantia Supertanker Parts," 3Dprintingindustry.com, January 22, 2018, https://3dprintingindustry.com/news/3d-printing-saves-17-navantia-supertanker-parts-127740.
4. Nyshka Chandran, "3D Printing Just Made Space Travel Cheaper," CNBC, June 7, 2015, http://www.cnbc.com/2015/06/07/3d-printing-made-space-travel-cheaper.html.
5. *The Economist*, "Print My Ride," June 23, 2016, http://www.economist.com/news/business/21701182-mass-market-carmaker-starts-customising-vehicles-individually-print-my-ride.

6. Joe Bargmann, "Urbee 2, the 3D-Printed Car That Will Drive across the Country," *Popular Mechanics*, November 4, 2013, https://www.popular-mechanics.com/cars/a9645/urbee-2-the-3d-printed-car-that-will-drive-across-the-country-16119485.

7. PwC and The Manufacturing Institute, *3D Printing and the New Shape of Industrial Manufacturing*, June 2014, http://www.themanufacturing institute.org/~/media/2D80B8EDCCB648BCB4B53BBAB26BED4B/3D_ Printing.pdf.

8. J. Hurley, "Loudspeaker Firm Sounds Out 3D Printing," *Daily Telegraph*, August 18, 2013.

9. Chris Perkins, "Porsche Has Started 3D Printing Parts for Classic Cars," *Road & Track*, February 12, 2018, https://www.roadandtrack.com/car-culture/classic-cars/a17045976/porsche-classic-3d-printed-parts.

10. EFT, *2015 3PL Contracting Report*, September 8, 2015, originally available at https://www.eft.com/technology/3pl-contracting-report-2015.

11. Aaron Bresinger and Jeff Arnold, "The Advancement of 3D Printing and Its Impact on Manufacturing and Distribution," *Supply & Demand Chain Executive*, July 7, 2015, http://www.sdcexec.com/article/12090442/the-advancement-of-3d-printing-and-its-impact-on-manufacturing-and-distribution.

12. WRAP, "The Value of Re-using Household Waste Electrical and Electronic Equipment," WRAP, accessed September 26, 2016, http://www.wrap.org .uk/content/value-re-using-household-waste-electrical-and-electronic-equipment.

13. Hans-Georg Kaltenbrunner, "How 3D Printing Is Set to Shake Up Manufacturing Supply Chains," *The Guardian*, November 25, 2014, https:// www.theguardian.com/sustainable-business/2014/nov/25/how-3d-printing-is-set-to-shake-up-manufacturing-supply-chains.

14. Andrew Allen, "Doubling 3D Printer Sales Herald Procurement as 'Core Differentiator,'" *Supply Management*, October 21, 2016, https://www.cips .org/supply-management/news/2016/october/doubling-3d-printer-sales-herald-procurement-as-core-differentiator-.

15. European Commission Directorate-General, "Explanatory Notes on the EU VAT Changes to the Place of Supply of Telecommunications, Broadcasting and Electronic Services That Enter into Force in 2015," April 3, 2014, https://ec.europa.eu/taxation_customs/sites/taxation/files/resources/ documents/taxation/vat/how_vat_works/telecom/explanatory_notes_ 2015_en.pdf.

16. Stephen Eisenhammer, "Oil Industry Joins World of 3D Printing," *Reuters,* January 23, 2014, http://www.reuters.com/article/ge-3dprinting-idUSL 5N0KW2OA20140123.
17. The Engineer, "How Local Manufacturing Is Redefining Humanitarian Aid," May 16, 2018, http://www.theengineer.co.uk/local-manufacturing-humanitarian-aid.
18. Sam Jones, "When Disaster Strikes, It's Time to Fly In the 3D Printers," *The Guardian,* December 30, 2015.
19. Kenji Ono, et al., "Analyzing and Simulating Supply Chain Disruptions to the Automobile Industry Based on Experiences of the Great East Japan Earthquake," *Journal of Integrated Disaster Risk Management* 5, no. 2 (2015), doi: 10.5595/idrim.2015.0102.
20. Dennis Fisher, "Japan Disaster Shakes Up Supply-Chain Strategies," *Harvard Business School Working Knowledge,* May 31, 2011, http://hbswk.hbs .edu/item/japan-disaster-shakes-up-supply-chain-strategies.
21. Amit S Mukherjee, *The Spider's Strategy: Creating Networks to Avert Crisis, Create Change, and Really Get Ahead* (Upper Saddle River, NJ: FT Press, 2008).
22. Sculpteo, *The State of 3D Printing,* 2017, https://www.sculpteo.com/en/ get/report/state_of_3D_printing_2017/.

CHAPTER SEVEN

Emerging Supply Chain Models

N THE MIDDLE of the nineteenth century, manufacture was predominantly local: things were made and bought largely within one's immediate area. If you needed something, you would venture around the corner to the local blacksmith, butcher, or wheelwright, and they would make what you needed. As mass production emerged, so did expanded global trade, and supply chains began to shift away from local areas. Manufacture was relocated wholly or piecemeal to other places where there were advantages to be had, such as cheaper labor, proximity to raw materials, or fiscal incentives. Those supply chains became truly global in the twentieth century and, by the end of that period, it wasn't just manufacturing that was sent far away. The end of the 1990s saw an explosion in offshoring other supply chains; service-based jobs in finance, human resources, and information technology were moved away in shared services and call centers that offered companies better cost and service models. It also became common for an item to be designed in one country, have its component parts manufactured in several others, have these shipped to and assembled in another location, and be sold in a totally different market. Those practices lengthened supply chains, making them more complex.

The early twenty-first century has seen a change in that trend: companies are looking to simplify supply chains, to bring their individual elements closer together, and as a result, local manufacture is now more popular. Industry is turning back to that time when more things were made closer to home. Several factors have catalyzed this change. On the demand side, many customers recognize that long supply chains mean long lead times. Moreover, they are also increasingly aware that those extended, complex logistics networks are environmentally unsustainable—an element of growing importance in customers' decision processes. For instance, consumer pressure has led supermarkets to identify the farms where their produce comes from.

On the supply side, many firms now recognize that long, complex supply chains present a high risk, with key elements exposed to significant natural and human-made disruption, from tsunamis and earthquakes to unstable governments and increased oil prices. In many of the countries where manufacturing was sent to because of cheap labor, local wages have risen and no longer offer an economic incentive. Complex, long supply chains are also difficult to manage, often lacking clear visibility. This makes it difficult to understand the state and levels of raw materials, work in progress, and finished products as they move from one part of the chain to the next, or to be able to make informed decisions to improve end-to-end efficiency.

Ultimately, whether global or local, supply chains are driven predominantly by three things: time, cost, and order accuracy. Getting the right product to the right place at the right time is the goal of every supply chain manager and the expectation of every customer. In parallel, those customers are constantly pushing prices down. They are also less willing to wait, expecting ever shorter delivery times. These factors are driving companies to evolve their supply chains, adopting the best models that meet their customers' needs while balancing their own costs and budgets to do so. For example, many of Amazon's customers want very fast deliveries, so to meet those needs, the company began working on solutions to deliver a select range of products in bigger cities within an hour of a customer placing an order. They achieved this by combining excellent predictive analytics to identify what is likely to be wanted in an area, considering many human, societal, and environmental factors; using delivery vans as mobile warehouses; and, where possible, having drones make deliveries. As mentioned briefly in Chapter 6, Amazon has gone so far as to patent "anticipatory shipping," whereby it will send products before they're ordered, knowing that there will be a demand for them and redirecting the logistics involved to make the delivery.[1]

The pressures that drove supply chains to become more widespread in the first place are ongoing, and it is in balancing these that 3D printing will have a significant impact on shaping them, such as by bringing production closer to home, thereby lowering costs and more closely meeting customer needs. For example, when customers need to get a critical, customized part on the same day for an urgent warranty repair, a relatively local 3D printer can meet that need. When customers need to replace a part that hasn't been made for 25 years because it comes from generations-old equipment or its original suppliers have gone out of business, 3D printing can provide a cost-effective solution. When a supply chain must reduce inventory to release working capital but also maintain service levels, 3D printing provides the necessary responsiveness. Where logistics are complicated, such as on an offshore oil platform or at a remote mine, again 3D printing is a solution. 3D printing is "front and center" when it comes to realizing local manufacture. Ed Morris, former director of mechanical engineering and manufacturing at Lockheed Martin, and today the director of the National Additive Manufacturing Innovation Institute, said of 3D printing:

> "[It] tears the global supply chain apart and re-assembles it as a new, local system."[2]

This has been noted at national levels: China, for example, has been investing heavily in 3D printing to reduce its reliance on the importation of parts and to keep costs low by eliminating the need for long logistics chains. In 2017 alone, the country saw US$ 1.1 billion invested in the technology.[3] This movement is part of what has been termed the "democratization of technology," where technology becomes increasingly accessible to more people and better able to deliver whatever is needed to whoever needs it, when and where they need it. It has been ongoing for 30 years in many areas of business and commerce, and now 3D printing promises to bring it to manufacturing and the wider supply chain.

When the word "innovation" is used in a conversation, the first thoughts are of some new technology, and usually that is the only form of innovation that people consider. Little thought is given to how it changes products or processes,[4] at least not initially. The biggest impacts of new technologies don't come from the tech itself but rather from how it changes companies' ways of working. Clayton Christensen, who helped originate the concept of disruptive technology, recognized that only rarely was it just the technology that was disruptive, that what actually made the big difference was how companies made use of it, opening up innovative operational models and supply chains,[5] something he termed "disruptive innovation." The advent of personal computers changed how supply chains collect, manage, and use data. More recent developments in sensor technology that, together with data analytics, broadband connectivity, and the Internet, have enabled the concept of Industry 4.0 are inspiring even more radical changes in the dynamics and operations of those supply chains.

The good news is that the disruptive nature of 3D printing is already driving some business leaders to change how they arrange and operate their organizations. This can now go beyond internal changes and involve entirely new supply chain models that are faster, more flexible, and more responsive than previously possible. *This is the most significant impact of 3D printing on supply chains.* The combination of being able to release designers from past constraints, to simplify designs, to make items in lots of one just as easily as lots of many, and the increased ability to deploy manufacturing more widely gives supply chains flexibilities and responsiveness that have never been achievable before, providing for a distributed "on demand" capability. It allows the boundaries between the supplier and the customer to be blurred, with manufacturing executed ever closer to the customer, potentially on their own premises, shortening supply chains and therefore response times.

The possibilities are exciting. For example, the metallurgical company Metalysis is aiming to eliminate the mining spare parts supply chain completely. The company, which produces metal powders for 3D printing from ores using an innovative electrochemical process, is piloting a model whereby its metal-powder production facility is sited at a mine in Kazakhstan whose products are used with locally placed 3D printers to make spare parts and manufacturing items for use in the mine.[6] This model is currently the preferred option being explored by NASA and other organizations that are planning for future lunar and even Martian settlements.

This chapter will first look at what it means to be a disruptive technology, how 3D printing fits that definition, and why that matters to supply chains. Next, it takes a closer look at the emergent supply models that 3D printing enables, grouping them in three sets:

1. In-house manufacture (IHM)
2. Customer-located manufacture (CLM)
3. Customer-managed inventory (CMI)

Each is described in terms of its characteristics and operation. These new operational frontiers will allow supply chains to meet their goals faster, more cheaply, and more accurately than has ever been possible, from improving operations to making step changes in performance.

A DISRUPTIVE TECHNOLOGY

The pace of change in technology over the last 150 years has been accelerating with each passing decade, and the list of technologies that have affected what we consume, how we use things, and how things are made is considerable. In that timeframe, humans have gone from only being able to get a few feet off the ground using hot air balloons to landing robotic probes on comets. City streets went from using oil lamps for illumination to automated light emitting diode (LED) bulbs, factories from steam to solar power. Communication has gone from horse-based to smartphones, and those devices give us access to more knowledge and data than people in the mid-nineteenth century would have been able to accumulate in their entire lifetimes. All too often, technological advances are said to be completely transformational or "disruptive." Sadly, like other terms used to described exciting things, that term is frequently misused.

The term "disruptive technology" was originally coined by the Harvard Business School professors Clayton Christensen and Joseph Bower in 1995.[7] They wanted to explain why companies failed to recognize when emerging technological developments arose that would go on to completely change their fortunes, such as Digital Equipment Corporation's (DEC's) failure to foresee the threat from the personal computer, or Bucyrus-Erie missing how hydraulic excavators would affect the demand for the big-bucket steam- and diesel-powered cable shovels offered by Caterpillar and John Deere. At the heart of this they found that those companies' customers focused on the technologies that they were accustomed to, at the expense of disruptive solutions.

When they looked at past examples of disruptive technologies, they found that when those emerged, mainstream firms didn't find them particularly useful or attractive: they didn't cater to the needs of their existing customer base, projected margins were too small, and adoption was seen as too complex and costly. Eventually, a new entrant brought the disruptive technology to a new customer set, and it became established there. That foothold on the market allowed for the development of the characteristics previously dismissed as not being good enough, enabling those firms adopting the disruptive technology to meet and surpass the needs of mainstream customers. DEC's customers were wedded to their large minicomputers. Bucyrus-Erie's customers wanted the raw power of their steam- and diesel-based machines. While both companies were comfortable, the disruptors were ignored, but they steadily made gains in performance, eventually taking those established companies by surprise. "Managers," wrote Christensen and Bower, "must beware of ignoring new technologies that don't initially meet the needs of their mainstream customers." They summarized that disruptive technologies display two common aspects:

1. They have a different set of performance characteristics that are not immediately valued by existing customers, doing better in areas that are less of interest to those, while underperforming at those that are.
2. The pace of improvement in the characteristics that customers most value is able to be met by the new technology sometime after it first emerges; essentially, it is born before its time.

Interestingly, the technologies themselves aren't always particularly new or advanced. They are often seen as quirky and of little interest, but over time, their superior capabilities become precisely those that are required to meet fast-moving customer needs. Consider the emergence of mobile telephones, from their early days as bulky equipment with patchy coverage, to today's smartphones, or the transformation of the music and movie sectors by the advent of the Internet and broadband technologies. Other examples of disruptive technologies include:

- Compact disks, which decimated the vinyl record sector
- Liquid crystal displays, which virtually killed off cathode ray tubes
- Digital cameras, which took over from film cameras

The aspects that Christensen and Bower identified perfectly describe 3D printing. It was initially seen as quirky, and was adopted by designers as a tool for prototyping, where the advantages in speed and flexibility of design outweighed the shortcomings in quality of finish, cost, and range of materials. Over the last 30 years, however, it has improved considerably, to the point where today it can complement, if not challenge outright, established, traditional manufacturing. 3D printing is, by that definition, disruptive.*

The question of whether 3D printing is disruptive or not is more than a game of semantics; it is an important label. Companies that recognize

---

* Christensen highlighted that there is a difference between "disruptive" innovations and those that merely offer their users an improvement on existing offerings, something that he termed "sustaining" innovations. This set contains the incremental advances in products, such as has been the case for the manual shaving razor for decades, and it sees the established firms in the sector continue to be so, such as Gillette in the shaving sector. Although 3D printing is undoubtedly a disruptive innovation—actually a set of different technologies—the advances in those technologies that companies like 3D Systems, Stratasys, and other leading firms achieve are thus "sustaining" innovations in themselves.

disruptive technologies and respond appropriately to them have a far higher probability of succeeding in the longer term and of surpassing their competitors. An example of this can be seen in the development of transistor radios. In their early days, these had poor-quality, "tinny" sound and were sold cheaply to teenagers. Development in electronics led to significant improvements in the fidelity of the sound without increasing the price, and sales of transistor radios overtook those of the expensive, higher-quality vacuum tube radios that previously had dominated the market. Very quickly thereafter, the older radio sets dropped completely out of the market, replaced by the cheaper, smaller, and far better transistor radios. The companies that saw the potential in the new technology and embraced the opportunity, like Sony, soon overtook the more established firms and dominated the radio market.

Following the success of disruptive technologies' original concept, Christensen identified that their biggest impacts came from how they affected operational models, something he termed "disruptive innovation." An example of this is the company Netflix, which started its life in the postal-based DVD rental market. It then saw the opportunity that the Internet brought, particularly the development of broadband, and used that as a platform to overtake rivals such as Blockbuster, which stayed wedded to its bricks-and-mortar model. From its founding, Netflix took less than 12 years to annihilate Blockbuster.[†8] Likewise, the biggest impact from 3D printing comes from how it is used to change operational models. Supply chains become truly supercharged, able to do things that they couldn't do previously. The biggest threat to companies that don't recognize the effect that 3D printing can have on their supply chains is that they will be overtaken by others that do, losing revenues and market share and, ultimately, failing. Just ask those traditional hearing aid makers that didn't adapt to the new reality of their industry.

▪ IN-HOUSE MANUFACTURE

3D printing enables a variety of innovative operations, and the most straightforward of them is described as in-house manufacture (IHM). Here, a company

---

[†] A few years before Blockbuster was forced out of business, Reed Hastings, founder of Netflix, at the time itself close to failing because of the costs of its mail-based operating model, called for a meeting with Blockbuster with the goal of the established company investing in the upstart. Blockbuster didn't spot the disruption that Netflix embodied, and Hastings returned home empty-handed. Blockbuster shut its last US store January 2014 in a year that saw Netflix earn US$ 5.5 billion in revenue.

adds 3D printing to the range of production methodologies that it already uses, or replaces some or all of its existing manufacturing with 3D printers. For instance, 3D printers can be used to produce the molds for its injection molding processes. This model provides the company with the benefits in flexibility that 3D printing affords, allowing more complex items to be made, lowering the cost of production through reduced wastage, and so on. The customers of those supply chains that adopt 3D printing in this way may well see an improvement in the responsiveness to their changing requirements. Many of today's examples of supply chains using 3D printing, such as GE's Leading Edge Aviation Propulsion (LEAP) engine fuel nozzle, are based on the IHM model, using 3D printing as a substitute for or extension of already existing manufacture.

To employ the IHM model, a supply chain first needs to identify which items may be produced using 3D printing in ways that bring benefits, such as lowering production costs, greater freedom in design, and so on. This calls for an understanding of the product and parts characteristics that are employed and made, such as their form and materials, lead times for delivery, costs, and so on. That data will also inform the choice of 3D printing technology and identify opportunities from redesigning items for 3D printing, such as simplifying or including design features otherwise not possible. Objects' designs must be digital: new designs can be so from the outset, while legacy items may need their designs converted into the right format. Once the business case demonstrates that 3D printing is the better option and the design data is available, a decision is needed on the best channel for obtain the services of a 3D printer, either leasing or buying one (Figure 7.1), or assigning the job to a 3D printer hub (Figure 7.2). From an implementation perspective, the IHM model is the quickest way to adopt 3D printing in the supply chain, potentially taking a few days to implement.

One common use for the IHM model is to drive down inventory levels that supply chains hold as they shift from an operational model that sees them making to stock or storing large numbers of spares and other items, to

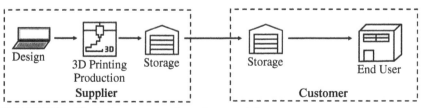

FIGURE 7.1 The in-house manufacturing model.

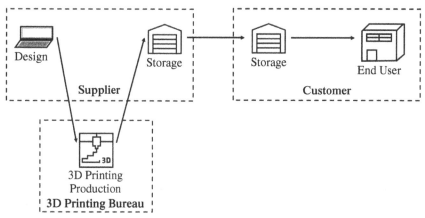

FIGURE 7.2 The in-house manufacturing model with outsourced production.

producing them on demand from digital files. This is the very embodiment of a make-to-order strategy, an example of which is the Deutsche Bahn case example described earlier in Chapter 5. Reducing inventory levels while moving to make-to-order can potentially reduce the necessary footprints of retail stores, since store-based stock is a key contributor to their size. Moreover, items made to order can be produced with the most up-to-date designs, incorporating amendments or modifications made since their initial design.

CUSTOMER-LOCATED MANUFACTURE

While the IHM model keeps manufacturing closer to the supplier, directly or indirectly with a 3D printing hub, the customer-located manufacture (CLM) model takes advantage of the greater flexibility in where to locate manufacturing, putting it where one needs it most (Figure 7.3). Typically, this involves the supplier placing 3D-printer-based manufacturing closer to, or even within, their customer's location, enabling the supplier to provide the customer with the parts more quickly. For instance, the customer identifies a requirement and sends a demand signal to the supplier, who in turn sends the digital file to the 3D printer near or at the customer's location, where the required item is then produced. In the right circumstances, that customer location can virtually be anywhere in the world. Here, the digital design file and the instruction of 3D printers remain under the control of the supplier, as does operating, maintaining, and repairing it. Issues such as postprocessing and finishing are

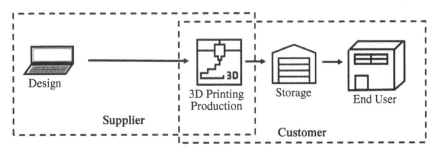

FIGURE 7.3   The customer-located manufacturing model.

the responsibility of the supplier and need to be included in the arrangement of the solution.

The earlier example involving the placement of Made in Space's 3D printer on the ISS is a good example of how a CLM model can work. The astronaut on the ISS as the customer sends a "buy" signal to NASA as the supplier, which then transmits the design data to the ISS-installed 3D printer to make what was required. This model can radically alter the dynamics of a company, and many firms are already actively investigating it, most notably those in the automotive sector, such as Ford, Mercedes-Benz, and Daimler, with the aspiration of placing 3D printers in vehicle servicing centers to produce spare parts. When a customer needs a part that can be made on the locally sited 3D printer, the manufacturer sends the design file and the part is made locally, rather than having to be shipped to the center. That is a particularly enticing proposition, particularly when logistics are difficult and distances large. In effect, like with the ISS example, delivery lead times drastically reduce, limited to the time to make an object. Amazon's recent patent to place 3D printers on trucks that can print en route to their destination, or be temporarily located at the end customer's base, is another example of CLM.

The viability of the CLM model depends on several factors, many of which are significant hurdles to 3D printing generally. First, as with the IHM model, it requires a good understanding of what items a company supplies, the materials that go into making it, how it is designed, and so on to identify candidate items for 3D printing. For instance, items are restricted in how many materials can comprise them, and they must require little or no assembly. Certainly, a key factor in the business case for CLM will be those lead times that customers face—can a CLM approach offer better times than would be otherwise the case? In situations involving international borders, overseas logistics, and geographical restrictions, CLM may well be the preferred option. The model only works if the entire production process for a needed item can be carried out locally, which requires the necessary equipment, raw materials, and operating skills to be locally available, as well as the capability for local postproduction and finishing. Other business functions have similar challenges, and lessons learned from those can be leveraged. For instance, many firms employ IT equipment that is located at their premises but provided and supported by a third party, such as PCs, server racks, and photocopiers. Those customers may carry out basic maintenance and repairs, such as clearing paper blockages or replacing ink cartridges on photocopiers, and the supplier carries out the more involved work.

One way to mitigate this need for local supplier personnel is to push the IHM model with outsourced production further toward the CLM model and use an outsourced production facility closer to the customer to obtain the lead time benefits. Already, companies like Fast Radius and 3D Hubs have witnessed several of their clients using this outsourcing-enabled version of CLM, with lead times reduced to next day.

CUSTOMER-MANAGED INVENTORY

Customer-located manufacture keeps responsibility for 3D-printing manufacture with the supplier, which responds to a buy signal from their customer and sends a digital file to the 3D printer near them. Customer-managed inventory (CMI) moves that responsibility to the customer. The 1980s and 1990s saw the emergence of vendor-managed inventory (VMI). With this replenishment model, a supplier is contracted to ensure that there is a minimum stock level of its products at the right place at all times, leaving the customer responsible for scheduling logistics, demand sensing, and so on. As opposed to the CLM model, where the control of what is produced and when it is made rests with the supplier, the CMI model (Figure 7.4) places that control with the customer, using 3D printing to make items on demand with little or no interaction with the supplier. Moreover, the customer themselves have their own 3D printers, owned or leased, with many potential technologies represented, giving them a wider range of materials to work with. Digital designs are available on demand through a licensing agreement, and the customer then instructs their own 3D printers to produce the desired object, much as they would with a local Just in Time dispensary. If the customer has the right 3D printers available, they can enter into CMI relationships with several suppliers, whose products all can be made on the same printers.

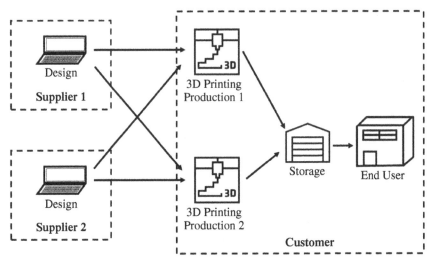

FIGURE 7.4   The customer-managed inventory model.

The CMI model holds promise for several sectors, particularly those with difficult logistics such as the armed forces. In 2013, the US Army deployed its Rapid Equipping Force (REF) to Afghanistan to function as a local "fablab," essentially a factory-in-a-box. The REF was equipped with 3D printers, Computer Numerical Control (CNC) milling machines, and other tools to produce spare parts locally far more quickly than resupplying by aircraft. A CMI model, which would remove the Computer-Aided Design (CAD) burden from the Army personnel by giving them locally available item designs, is being investigated as a potential future development. As James Zunino, a materials engineer for the US Army Armament Research, Development and Engineering Center, told a conference on the technology in the military environment:

> "If you are a soldier in a FOB (Forward Operations Base) in Afghanistan, everything is different. It's not as easy as running down to the Home Depot and picking up a screwdriver. Those costs add up, when you add all the transportation costs, fuel, security, it then might be cheaper to be able to print one."[9]

The CMI model gives the customer far greater control of what items they have and can make at any given moment, taking away the supplier's involvement in those. Delivery lead times are reduced to the time to make the items.

The CMI model is the most complex of these new emerging models, and while currently no companies are actively using it, many are contemplating it, held back by the wider considerations. One of these, explored in more detail in Chapter 8, is the need for suppliers to develop new pricing models to accommodate 3D printing. This is particularly important for the CMI model, where suppliers are virtually distant from manufacture and delivery supply chain activities. There needs to be an effective licensing mechanism to ensure that the customer isn't abusing the model, such as by making 10 items while paying for one. Similar issues were faced by the music industry when digitally recording CDs become available to consumers, opening the possibility of mass piracy and threatening royalties. The solutions that that sector developed will no doubt be considered by those using 3D printing designs. As well as license controls, configuration control of the designs is also required, with suppliers having to ensure that customers have the latest approved digital designs for their specific requirements. The use and nature of digital designs make it easier to keep up to date, minimizing delays in production as well as procurement costs. Furthermore, as for the CLM model, issues of data security need to be considered.

TOWARD A VIRTUAL 3D-PRINTED SUPPLY CHAIN

Many of today's most innovative companies have utterly changed the conventional wisdom of what is needed to operate in their sectors. Uber is one of the largest companies in the transport sector, yet it owns no vehicles. Airbnb is one of the biggest companies in the hotel industry, yet it doesn't own any hotels. Both have done this by outsourcing those physical parts of their operations. This modus operandi is now entering manufacturing, with the beverage maker Fever Tree as a pioneer.

Fever Tree produces mixers such as tonic water, ginger ale, and colas. The company has had meteoric growth, something that had been particularly difficult to achieve in the beverages sector where other firms like Coca-Cola, Schweppes, and PepsiCo dominate worldwide. And yet the firm's growth since its founding in 2005 saw it achieve revenues of US\$ 225 million by 2017, with global year-on-year growth above 60% and a 2017 market capitalization above US\$ 3 billion. This has been achieved by outsourcing most of its manufacture and logistics to third parties, retaining a small team of full-time employees to focus on marketing, sales, quality control, finance, legal, and sourcing. The latter function has been critical to the success of their products, while a strong branding message has taken Fever Tree into most of the bars and pubs in the UK before opening up the US market.

Away from the beverages sector, this model of outsourcing the physical parts of the supply chain, from design through manufacturing to logistics, is particularly suited to supply chains that employ 3D printing. The benefits are clear:

▪ Fast start- and ramp-up times. As manufacturing is 3D-printing-based, it can be established far quicker than with traditional techniques. Moreover, it can be scaled as order volumes increase, recognizing the limitations in volumes and speed.

▪ Agility and responsiveness to market demand changes. 3D-printing-based manufacturing is able to respond to dynamic or customized needs given the greater freedoms in design that 3D printing offers and the ease with which designs can be altered.

▪ Lower logistical costs with higher service levels. As manufacturing can be located closer—or at—customer locations and offered as a service to produce items on demand, logistical costs are greatly reduced while retaining the capability to provide what is needed relatively quickly. Moreover, this can be accomplished across national borders and overseas, significantly reducing lead times and costs from tariffs and customs.

As in the case of Fever Tree, what is left for the company is marketing, sales, quality control, finance, and legal with the supply chain focusing on sourcing and supplier management (suppliers being the third-party designers, 3D printing hubs, and logistics companies). At the heart of such an operating model are issues of quality and IP, and not all firms are comfortable outsourcing what can be core competences to third parties. However, the advent of more widespread and high-quality 3D printing hubs, together with improved access to the skills needed to engender a Design for Additive Manufacture (DfAM) approach in designs, means that this model will emerge in the near future. It may come from existing supply chains that pivot their ways of working, as Deutsche Bahn is doing through the Mobility Goes Additive consortium. Alternatively, it will emerge from a totally new start-up. Wherever it comes from, the new model will inevitably upend the dynamics of the sector it is in.

The new operating models that 3D printing allows for will no doubt change how customer demands are met, offering companies innovative ways to reduce order times and increase customer service levels. Given the reasonable ease with which 3D printing can be added to or replace existing manufacture that we saw in the previous chapters, it is tempting for companies to dive right in and start 3D printing parts tomorrow. However, as with most strategic technologies, it isn't that simple—there are many wider implications that need to be taken into consideration, and many of those need to be planned for and changes made to how a business operates. These will be considered in the next chapter.

Notes

1. Joel R. Spiegel et al., "Method and System for Anticipatory Package Shipping," US Patent US8615473B2, filed November 28, 2011, and issued December 24, 2013.
2. Chuck Intrieri, "The Impact of 3D Printing in the Supply Chain and Logistics Arenas," Cerasis, February 10, 2014, http://cerasis.com/2014/02/10/3d-printing-supply-chain/.
3. Ralph Jennings, "China Lays Groundwork For Asian, World Lead In 3D Printing," *Forbes*, February 1, 2018. https://www.forbes.com/sites/ralph jennings/2018/02/01/china-lays-groundwork-for-asian-world-lead-in-3d-printing/.
4. Thierry Rayna and Ludmilla Striukova, "The Impact of 3D Printing Technologies on Business Model Innovation," in *Digital Enterprise Design & Management*, ed. P. Benghozi, D. Krob, A. Lonjon, and H. Panetto, vol. 261 (Berlin: Springer, 2014), 119–32.
5. Clayton Christensen, *The Innovator's Solution* (Boston: Harvard Business School Press, 2003).
6. Interview with the author, March 2018.
7. Joseph L. Bower and Clayton M. Christensen, "Disruptive Technologies: Catching the Wave," *Harvard Business Review*, January–February 1995, https://hbr.org/1995/01/disruptive-technologies-catching-the-wave.
8. Big Think Editors, "Ken Auletta on How Netflix Killed Blockbuster," April 3, 2014, http://bigthink.com/think-tank/ken-auletta-on-how-block buster-killed-netflix.
9. IQPC, "Additive Manufacture for Defense and Aerospace conference," Washington DC, 2015, http://additivemanufacturingfordefense.iqpc.com /media/1001723/46459.pdf.

CHAPTER EIGHT

Wider Implications of 3D Printing

T IS A fact that, with every invention, there is a use or consequence that its creator didn't expect and that had to be controlled. Cassette recorders were useful for recording voice and music, but legislation had to be amended to allow for the widespread tape-to-tape recording that they permitted, which conflicted with the payment of royalties. USB memory sticks are useful for transferring files between systems, but companies had to curb their use by imposing IT security measures to counter the threat from viruses and security breaches. 3D printing is no different, and there are several wider implications that supply chains, the companies involved in them, and others beyond those need to consider. Of course, not every supply chain will be able to control all those wider aspects, but they can and should take account of them.

The adoption of 3D printing would appear to be straightforward, and initially many will view it as simply a new way to make things, much as they would an innovative way to mill material or an improved lathe. However, 3D printing is a significantly different manufacturing technique, involving a radical change in how businesses operate, from design to distribution, and several implications need to be taken into account when considering or actually using it. These can be grouped into nine categories:

1. Legal, liability, and intellectual property (IP)
2. Quality management
3. Standards
4. Regulation and accreditation
5. Health and safety
6. Data management and security
7. Commercial models
8. Fiscal and financial implications
9. Skills base

This chapter will take a closer look at each of these, not by providing a deep technical analysis of each one together with a comprehensive set of solutions—for those, the reader is encouraged to seek out more specialized authors—but rather by describing the salient points that supply chain decision makers need to consider. At the end of each section, it will also offer recommended actions that decision makers can take to mitigate each issue. Many will be familiar to those who have previously considered and adopted other new technologies—after all, 3D printing is one such innovation. For instance, the issues described relating to the legal, liability, and IP topics are valid for any manufacturing process based in digital data models, such as computer numerical control (CNC) machining. However, the breadth of those issues and their significance in successfully leveraging 3D printing in supply chains are new. Without understanding these wider implications and then developing approaches to deal with each of them, stakeholders risk making decisions at the wrong time, or in the wrong way, essentially sentencing attempts to use 3D printing, whether in pilot projects or full implementations, to failure.

LEGAL, LIABILITY, AND INTELLECTUAL PROPERTY

Out of all the wider issues that 3D printing raises, the most critical and complicated are those related to the law. At the heart of those is the fact that 3D printing transfers legal risks away from the physical products and onto the information that goes into making those products. That shift directly affects decisions on how to use 3D printing in supply chains, while raising questions of what can be made using it. Should a company 3D print something whose very design is where its biggest value rests? What can be protected and how?

What does having 3D-printed items do to the legal exposure of the company? Can a supply chain opt to 3D print locally given the region's legal frameworks?

The flexibility, versatility, and openness of 3D printing challenge legal protections and systems and, as has been the case with other technological developments, both of these must evolve. The core issue is that 3D printing allows anyone with the right equipment to reproduce a design and make an existing object. According to the author Scott Grunewald:

> "3D printing technology is going to completely alter copyrights, trade-marks and IP law dramatically over the next few years simply because there really are not a lot of ways to stop people from duplicating, and in some cases stealing and taking credit for, 3D content."[1]

One approach to protect designs from illicit 3D printing has been to enforce blanket bans. For example, until 2016, the Kingdom of Jordan prohibited (but never banned outright) the import of 3D printers into the country, following the publicity in 2011 concerning 3D-printed handguns in the US. After that date, companies wanting to use them could only do so under heavy regulation, which required the registration of users, the strict control of printer sales, and the use of cameras inside shops selling them.[2] This is perhaps a step too far for most and it certainly has not been the norm everywhere. The legal issues still present wider implications to organizations, which can be grouped into two areas:

1. IP
2. Liability

IP

The changing nature of where legal risks lie means that the need for IP protection in 3D-printing situations is augmented. As the quality of the 3D-printed items improves, as times to manufacture them and the costs involved shrink, the appeal of using the technology to produce counterfeit goods will inevitably grow. The threat is significant, with some authorities estimating that more than US\$ 100 billion per year will be lost in IP violations globally.[3] Despite the scale of this threat, IP protection for 3D-printed parts is vexing many lawyers in the field, and chief operating officers, supply chain managers, and designers are now beginning to consider the issue more seriously. Naturally, designers want their work to be covered; corporations want their revenues to be protected; and customers want the assurance that what they buy is safe, that it comes from where they expect it to, and that it will act as intended, with the guarantee of the advertised quality. The 3D-printing ecosystem, though, threatens all of those, making design piracy and the counterfeiting of goods far easier than previously possible. As has happened with several technology developments in the past, most notably in the music and film sectors, 3D printing will inevitably have a "Napster moment," the point at which reproducing protected designs is so widespread and difficult to control that existing IP laws are no longer enforceable. So, what can be done about this?

To answer that, it is first necessary to have a basic understanding of exactly what IP law is. It exists to protect innovations and creative works, in the form of copyrights, designs, patents, and trademarks:

- Copyright refers to the unregistered protection of a creative work, such as an artwork or sculpture, a song comprising musical notes and lyrics, or a story.
- Designs, which can be registered with patent offices or not, refer to protection of how things look—their form.
- Patents, which are registered with a patent or IP office, protect how something works.
- Trademarks (and trade names), which can also be registered, are a "badge of origin," providing assurance that what is being protected is genuinely from where it claims to be.

An object will normally have its look or form protected by a registered or unregistered design; how it works and its function will be protected by patent, with the "brand" providing assurance that it comes from the expected source through a trademark or trade name. If there is a change to the form of the part, then a new design protection is needed. For instance, this would be the case if redesigning a part as a single unit rather than one assembled from many components. If the design changes how it functions, how its internals work, or how it is used, then a new patent may be needed, such as if a robot arm were redesigned to have fewer parts by internalizing the hydraulics.

So how does 3D printing threaten these different forms of IP protection? IP law is notoriously complex when dealing with traditional, existing technology, and new technology adds to that complexity. In the first instance is the question of whether any breaches of IP, whether copyright, designs, patents, or trademarks, are carried out by the manufacturers of 3D printers directly or even through their provision of printers to those who commit the breach. Can Stratasys, 3D Systems, or HP, for instance, be held liable if someone duplicates and produces items from a protected design using their machines?

To answer that, IP lawyers point at past case law under similar conditions, such as the case *Sony v. Universal City Studios* in the USA in 1984. In this case, which became known as the "Betamax Case," Sony, as the manufacturer of Betamax and VHS machines, was found not liable for the use of its machines in situations of copyright infringement. A similar case in the United Kingdom in 1988, *CBS v. Amstrad*, ruled that Amstrad was not liable for copyright infringement when its twin cassette machines were used to copy cassettes. Based on those cases, the manufacturers of 3D printers would not be held liable for IP breaches of 3D printing designs manufactured on their machines. However, it is worth pointing out that many national jurisdictions don't rely on legal precedent; in those situations, the makers of 3D printers may still be found liable. It is quite likely, then, that printer manufacturers will limit authorized sales of their equipment to those regions where they can safely avoid liability, and this restriction needs to be considered when planning where to place printers in supply chains.

What about other steps in the 3D printing process? As we know from the experiences with Napster, digital files are easily copied and distributed, and tying down those responsible for having obtained the files in the first place is not easy. Moreover, in today's connected world, there is a general acceptance of open-source materials whatever their provenance, which raises the tolerance of digital files that aren't obtained in an orthodox, legitimate fashion. This means that the sharing of files already "out there" is more likely. Therefore, there needs to be more restrictive control of design files beyond that typically found in traditional environments, only allowing access to authorized users for limited purposes. This can be achieved through encryption, access control, digital signatures, or other technical methods. However, as IT security experts have known for decades, the weakest point in the chain will always be the human, so control processes and procedures are also required. According to identity security specialist Andrew Hindle:

> "3D printer manufacturers also need to be part of the solution. For example, it should be possible to encrypt and digitally sign design files, possibly through CAD packages. 3D printers therefore require the capability to decrypt and verify the digital signature. Moreover, they should be configured such that only 'signed' files can be printed."[4]

With the right technical tools, such as blockchain, these signatures can provide assurance of design origins, allowing for configuration control and protecting them from illicit use and tampering.

Direct IP infringement is difficult to enforce, though, as has been the case in several other technology sectors involving digital media. However, those sectors have indicated that the focus should be on using indirect infringement instead. When the music industry was faced with song tracks being reproduced and shared openly on websites such as Napster, and the film industry likewise saw entire movies being shared on sites like Pirate Bay, it was those third-party sites themselves that were prosecuted rather than the individuals who supplied the original data. In these two case examples, Napster and Pirate Bay were found liable for indirect infringement. Likewise, 3D printing designs are today shared on dedicated and general data sites; if a protected design is hosted on a site, then it will be that site that is prosecuted for IP infringement.

As well as designs, the protection of IP affects manufacturing itself, primarily with regard to quality assurance (QA). When a customer purchases an item from a supplier, trademarks and brands indicate that the part is legitimate and that it has undergone the requisite QA that the supplier has in place. The difference between a legitimate part and a counterfeit one is often the attention paid to quality, and brands—particularly highly specialized, luxury, and high-quality brands—differentiate themselves based on that strong level of assurance. This is especially so in those sectors that employ parts that are safety critical, such as the aerospace, defense, and automotive sectors. It also affects those brands where there is inherent value in the brand deriving from the perceived quality of the object. In the jewelry sector, the perception of a piece's quality and the degree of workmanship needed to make it translate into high prices. Today, to judge by the websites of 3D printer makers, many high-end jewelers employ the technology but not all admit publicly to doing so, perhaps for fear of bursting the perception bubble that justifies their high prices. If the value doesn't derive from the manufacturing process, then it must stem from the design, and that then needs to be protected. Moreover, when a customer buys a part that has been made on the same 3D printer as an original, set up and maintained in the same manner, using the same design as that original, there will be a question of whether the finished part is of the same quality as the original. If a trademark is included in the design, then it will be assumed that this is the case. Designers therefore must be careful to brand designs and place trademarks on them only at the right point in the manufacturing process, considering where they can provide the necessary assurance during fabrication.

As in the cases mentioned earlier, IP law tends to evolve based on case history, and IP lawyers are continuously and closely monitoring 3D-printing-related cases to help prepare for additions and amendments to the current body of law. In the meantime, what can companies do now to prepare for the issues that will arise? In the short term, companies considering

3D printing will have to evaluate their current IP protection, asking what specifically is protected by designs, patents, and trademarks, and whether those protections are registered or unregistered. They should look at whether the protection covers entire parts and products, or the components to assemble those. Next, understanding what can be 3D printed will help to answer the question of what will need protecting and, from that information, firms can assess whether current IP protection applies or whether new solutions are needed, thereby helping to define what will be 3D printed. If there is doubt, then "patent, patent, patent"—considering designs, techniques, and anything else that can be protected—and ensure that 3D printing is covered in the scope of the patent application. In parallel, protection measures can be built by leveraging security innovations to resolve some of the issues. For instance, counterfeiting protection may be achieved by embedding an encrypted trademark within the physical structure of a part. Texas Quantum Materials Corp., for example, licenses "quantum dots," tiny nanocrystals that reduce counterfeiting by giving products a "physically unclonable signature" that only the manufacturer knows, and other technological solutions are fast emerging.[5] Keeping the making of 3D-printed prototypes in-house can also help reduce the risk, but ultimately designs can be reverse-engineered, so other mitigation actions are needed, particularly data management (see "Data Management and Security" section in this chapter).

Liability

James Beck, a lawyer with the firm Reed Smith, calls the question of liability and 3D printing "a wild, wild West" legal issue.[6] At the heart of this, he says, is one fundamental question:

> "What happens to product liability law when users become manufacturers?"

Imagine the situation where someone 3D prints an item, installs it, uses it, and the part fails—who is liable for that failure? The user who employs it, the installer who fitted it, the maker who printed the part, the printer manufacturer, or the original designer? Answering this question (Figure 8.1) requires a long investigation into what happened, just as any other industrial incident of this type would, asking about whether the design was sound; did the manufacturer use the right equipment and did it perform as expected; and was the part installed correctly, and subsequently used appropriately. Even so, the answer isn't clear-cut, requiring clarification of tort and the definitions

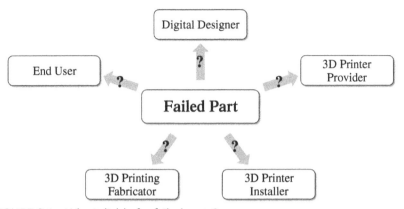

FIGURE 8.1   Who is liable for failed parts?

concerned to answer who the manufacturer is, who the designer is, and so on. US law, for instance, requires that an item have been "sold" in order for product liability laws to apply, something that isn't easy to qualify in many 3D printing situations.

3D printing opens new questions of liability in areas such as the availability of designs. Now that anyone can make any design using the right equipment, who is liable when a design file is used inappropriately? For instance, what happens when an authentic design for a part is used by a third party in breach of IP? Consider the situation of a toy company releasing designs for small parts that can be printed by a distributor, such as a toy arm or a shoe for a doll, and after printing, a child is injured by choking on that part. Who is liable? The current legislation covers many areas of product safety; however, it is probable that it will have to evolve to encompass the use of 3D printed parts, at home and in industry. In the meantime, licenses for the use of designs will most probably offer some measure of risk management.

The question of liability also refers to the management of designs for parts and equipment used for nefarious reasons. For instance, the USA has strict export regulations on the trade in weapons and equipment that can affect national security, such as parts that can be used to create improvised explosive devices (IEDs) and cryptographic equipment. Who is liable if the design of a 3D-printed rifle, either for the whole weapon or for component parts, become available on the Internet, or someone releases the design of a part that, although legitimate and designed for legitimate use, potentially can be used as a component for a bomb? This question was stirred up most prominently in the well-documented case of a 3D-printed handgun produced by an American organization, Defense Distributed. In 2013, this company released a video on YouTube demonstrating how it was possible to create a working handgun using a readily available 3D printer, in this case a Stratasys Dimension SST, which uses a plastic STL process. Named the "Liberator"—after a World War II weapon that equipped agents working in the Office of Strategic Services, a precursor of the Central Intelligence Agency—the handgun was composed of 16 individual, 3D-printed parts and a traditionally made metal firing pin. It was publicly tested at a firing range, successfully discharging a small number of rounds, with the video of the test garnering numerous views. Under current US legislation, the manufacture of the weapon was entirely legal, as it didn't contravene the National Firearms Act as an automatic weapon, for instance, and it would only be in breach if the company were to sell the weapon commercially without a license. However, the question of liability and legality becomes key if the designs were to be released, such as through a design-sharing platform. Indeed, this was something that the owner of Defense Distributed, Cody Wilson, initially did, and the design was downloaded over 100,000 times before it was taken down from the platform where Wilson hosted it under pressure from the US State Department, which threatened to prosecute him and his organization under the regulations prohibiting the export of unapproved arms. In response to that, Defense Distributed filed a lawsuit against the State Department in 2015 alleging several breaches of their Constitutional rights, but the case fell through. If a third party were to use those designs to make a weapon that was then used in a criminal case, the issue of liability would inevitably have to be tackled, and it is doubtful that current legislation would be sufficient to deal with it consistently.

During the Defense Distributed debate, the manufacturer of the 3D printer, Stratasys, opted to enforce the leasing agreement that they had with Wilson. They removed their equipment from his premises, alleging that the agreement specifically prohibited the machine from being used for illegal purposes, although the exact nature of those has not been identified publicly. It is likely that manufacturers of 3D printers will include the issue of liability in their terms of sale and lease, aiming to absolve them of any responsibility if the machines are used for illegal purposes, and it will be up to the national courts to test the strength of those claims against the manufacturers' duty of care.

Key Actions for Supply Chains

1. Review what needs to be protected before embarking on a path toward 3D printing. Consider what is already protected and why. Examine how that protection is enforced and review the mechanisms already being used to ensure that 3D printing is specifically covered.
2. Consider approaches to protect designs, techniques, and so on. This should include legal methods, such as patents, trademarks, and registrations as well as physical solutions such as embedding identifiers within designs.
3. Conduct liability risk management: identify threats from illicit copying and use of designs, then build contingency and mitigation plans accordingly.
4. Watch IP legislation carefully as it evolves to take due consideration of 3D printing.

 ## QUALITY MANAGEMENT

It should go without saying that there is a need to provide quality control (QC) and QA in all manufacturing industries. Both are key elements of the Make step of the supply chain model as they affect both the cycle times and costs of production. The difference between them is subtle but significant, with the International Standards Organization (ISO) 9001:2015 standard defining them thus[7]:

- QC is "part of quality management focused on fulfilling quality requirements."
- QA is "part of quality management focused on providing confidence that quality requirements will be fulfilled."

Ostensibly, then, QC refers to the set of activities that identify faults in something that is made—in its geometry, its finish, and the properties of the materials from which it is made—whereas QA refers to the set of activities that ensure the processes that produce what is made do so in the manner that they should. QC asks questions such as:

- Has the machine been set up correctly?
- Is the size and shape of the item correct?
- Is the surface as smooth as it needs to be?
- Is it as stiff or supple, strong or breakable, or long- or short-lasting as intended?

On the other hand, QA asks whether the right steps, procedures, and processes were followed, thereby encompassing QC and beyond. It ensures that there is consistency in production, that consecutively 3D-printed parts using the same design all have the same quality, even if these are produced at different times. Both QC and QA are needed to ensure that what is manufactured is precisely what was intended, and every traditionally manufactured item must undergo some form of both at some point to ensure that it is what it was intended to be. 3D printing is no different.

The maturity of traditional manufacturing techniques means that there is a plethora of proven quality management methods that those can employ. These include destructive (i.e. what gets tested is damaged or destroyed in the process) and nondestructive techniques, the use of inspection, and audits. They tend to be extensive and not only increase the cost and time to make things, but many of those also eliminate the very advantages of 3D printing that make it desirable in the first place. Imagine having to make one of every unique design variation of something so that it can be destroyed for QC and QA purposes. Fabricating things with 3D printing mandates new QC and QA paradigms, and these are only now beginning to mature. This is primarily because 3D printing technologies have only relatively recently moved from being used to make prototypes to producing finished items, and those different purposes call for different types of QC and QA. The situations where 3D printing will be used also present several challenges:

- The wide range of 3D printing technologies, each of which requires a different approach to QC and QA
- The production of "lots of one," meaning that every item needs to undergo QC and QA slightly differently
- The dislocation of production from designers and potentially those who "own" that production, such as when 3D printer hubs are used, or with the customer-located manufacture (CLM) and customer-managed inventory (CMI) supply chain models described in the previous chapter, with production in multiple locations, operating in multiple environments.

Consider the challenge of a car manufacturer's distributed spare parts concept, which calls for several items to be made in a range of materials, each needing a different 3D printing technology, if not a different printer, with those printers placed in multiple locations and operated by a variety of people. How does the supply chain carry out QC and QA in a repeatable, effective, and efficient manner with so many variables? For 3D printing, that extends to how the printer is set up, the position and orientation of every part to be produced, the angle of printer heads, the calibration of the postproduction and finishing machinery, how parts are moved through the process, and many other factors.

Another advantage of 3D printing is that it is as easy to make a complex part as it is a simple one, the so-called complexity is free benefit. Parts made with complex designs and complicated internal structures still need to be checked once they are made to ensure that they really are what was originally intended. Internal pathways and ducts for hydraulics or airflows, the thickness of internal support structures, and the clearances between surfaces all need to have some level of QC to ensure that they are as they should be. The extreme case of a 3D-printed liver, something that some firms are postulating, illustrated the scale of this task. Each liver has tens of thousands of internal channels through which blood will flow, and a blockage or deviation from design parameters in even a few of those can have serious consequences. So how does one test that the diameter and location of all those channels are correct? How can something with such complexity be checked when normal testing protocols are not suitable?

If a part fails QC and QA, there are the matters of who is responsible for the fault and remedying it. Identifying where the failure originated—the design, the 3D printer, the printing process, the materials used in manufacture—is the first step. Therefore, the best approach is to first break the end-to-end 3D printing process down into smaller parts, tackling each one in turn, before taking a holistic perspective. This requires focusing on each of the 3D printing process steps and understanding the extent that QC and QA is needed for it—not every item will be as complex or as critical as a liver. For instance, the end-to-end process can be divided into three areas (Figure 8.2):

1. *Design, modeling, and simulation.* The current design processes involving modeling and simulation already provide a good measure of QC by identifying faults and deviations from the intended parameters. Increasingly, the software tools to conduct these processes consider the particular

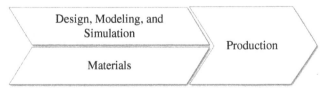

FIGURE 8.2 Framework for considering quality control and quality assurance.

nuances and limitations of 3D printing technologies, although designers and those managing the tools need to be clear on what technology and specific printer will be used, and their capabilities and limitations, much as they do with traditional manufacturing processes. Complexity, it turns out, is only free-ish, and there will be some things that can't be done. QA is then achieved by ensuring that the right steps are carried out, that designs go through the necessary checks and pass through the required modeling and simulation before being made ready to be fed into the next phase.

2. *Materials.* QC and QA are not just activities for the actual design and production phases, but also extend to all the feeds into those. In the case of fabrication, that starts with the materials themselves. When those are sourced, they naturally need to undergo QC/QA. The processes for handling and loading them into the 3D printers also expose them to potential "faults." Consider the powdered materials used in Fusion Deposition Modeling (FDM), for instance. Humidity, static, and movement all affect the nature of those, resulting in clumping and appearance of powder balls. When the powders are subjected to vibration, such as during transport, the clumps move to the top of the container that they're in—much like raisins rising to the top in boxes of cereal mixtures—and those affect how the powders act in the FDM process. QC processes and QA of those will minimize the risk of such issues.

3. *Production.* The actual 3D printing process being used can introduce flaws. For instance, in extrusion-based technologies, the nozzles that expel the raw materials at high temperatures and pressures will be affected by those conditions, degrading each time they're used and running the risk of clogging. This affects the tolerances and precision that they produce, cumulatively changing the characteristics of fabricated items; the tenth or twentieth item produced may not be of the same quality as the first item. Similarly, Powder Bed Fusion machines initially tended to produce holes or weaknesses in the fused materials, weakening the final object. Once an item has been 3D printed and it undergoes postproduction and finishing, it again needs to undergo QC to ensure that those final processes have been applied correctly. Lastly, QA confirms that printing, postproduction, and finishing have been carried out correctly.

While this sounds straightforward, current QC processes are simply not good enough to cope with the needs of 3D printing. Many QC checks are carried out manually, which is both slow and laborious, and are not sufficiently refined to catch all faults that can arise. A cybersecurity team led by New York

University professor Ramesh Karri tested the threat to 3D-printed objects posed by deliberately inserted faults. The sub-millimeter defects that the team placed into the example objects were not detectable using the current techniques, such as ultrasonic imaging.[8] Furthermore, quality-assuring 3D printing is made harder because today's QA techniques predominantly rely on statistical methods, and the population of parts from which to base the statistical analyses simply isn't sufficiently large to provide a reliable judgment, particularly when what is being made is one of a small number of items or a single unit. One method that ameliorates this lack of data is to group designs and objects into families with similar characteristics—for instance, those with a similar form or material—and define QA parameters at the family level.

The ultimate aim is to continuously perform QC during manufacturing processes, which are themselves constantly quality assured, and feed those results back into the 3D-printing processes so that corrective action can be taken; this is the basis of a closed-loop quality management system. To a large extent, this is what takes place in the design stages, where faults and deviations detected in the modeling and simulation tools are fed back to the designers, who make the necessary changes.

Embedding such approaches in the production phases has some practical hurdles. For instance, QC and QA methods need to be quick to carry out in the production phase, something that is especially important in situations where 3D printing is being used to provide mass customization, one of the biggest potential advantages of the technology and a driver for its increased use in supply chains. As yet, most 3D printers don't have that feedback loop. For now, therefore, QC and QA of 3D printed parts must rely one open-loop approaches, where checking is done manually after printing.

Of course, the basic elements of traditional quality management continue to apply to its implementation in a 3D printing situation, including ensuring that personnel are adequately trained, software correctly installed, and machinery properly calibrated. However, further development and innovation in this area is needed if supply chains are to have confidence in the QC and QA of 3D printed items. New approaches are already emerging. Efforts are underway at several university and government research departments to build up the necessary understanding of how 3D printing processes change material properties to better inform and shape QC and QA methodologies. Technology companies that make 3D printers like EOS and Additive Industries are increasingly building QC mechanisms into their machines to provide seamless and integrated quality management processes that reduce the need to pause fabrication or handle products manually and speed up cycle times.

This is achieved by using internal sensors and monitoring software that detect deviations and faults, and so ensure a predictable and repeatable process. For example, the company Sigma Labs offers designers and manufacturers an integrated, modular solution set, PrintRite3D, which "combines inspection, feedback, data collection and critical analysis."[9] This set uses a series of sensors—such as accelerometers, ultrasound sensors, and high-speed, high-resolution, and thermal cameras—and a range of data analytics to provide "objective evidence of compliance to design intent" and early fault detection. The practicality of those is limited by the scale of data produced and the processing power needed to carry out those analyses, and the advent of automation promises to speed up the QC and QA processes, such as the use of robotics to carry out machine visual metrology of every part made as it moves to and through postproduction.

Alongside those efforts, much research is looking specifically at the QA of 3D printing processes. Concerns about how 3D printing techniques affect the materials being used during fabrication are continuously being investigated by public- and private-sector researchers. Questions are being studied regarding the internal strengths of 3D printed parts; how the application of directed energy, such as in laser sintering, affects the properties of powders; how surface tensions in liquefied metals translates into the properties of the solidified material; and how the way heat dissipating from where energy beams are focused changes metallurgical structures. The outcomes of those inform the QA of processes as a whole as well as of specific manufacturers' 3D printers. The US Navy's Office of Naval Research, for instance, is leading efforts to investigate microstructures and predict material properties, including stress and fatigue, in metal 3D printing, using those to predict how items will perform. The next steps are to optimize the build approach for making metal objects and then to accelerate their acceptance.[10]

Much as the choice of 3D printing technology will depend on what is designed and how it will be used, so those will also inform the level of assurance needed. In the same way that prototyping doesn't necessarily require the highest levels of QC and QA, not every finished item will need to achieve the same level of quality. Of course, in situations where the integrity of a part is crucial, such as for safety-critical or high-spec items, then there is no question of being thorough; in most of those situations, regulation and accreditation will guide the required levels. However, that set of items is only one in the universe of 3D printed items. The consultancy Deloitte advocates a multi-tiered approach, which they have termed the Additive Manufacturing Quality Pyramid. This model arranges and articulates what is needed in a logical hierarchy,[11] identifying the necessary QC and QA for each situation. As the authors of that excellent and insightful paper state:

> "Quality will not—and should not—look the same for every type of part and product."

At the high end of the scale are the quality expectations for, say, high-precision tools, dental implants, or aerospace components, which can require tolerances in the micrometer (10^{-6} m) range, while at the other end are the millimetric (10^{-3} m) tolerances of concrete structures in civil engineering. An item needed for one-time use over a short period of time, such as a temporary plastic cover for a junction box, doesn't need the same durability and fatigue profile as a mount for a Trent 1000 jet engine. Quality is, as the saying goes, a case of "horses for courses."

How does this impact supply chains? The management of quality and the constraints that QC and QA have in a 3D printing environment mean that supply chains need to be aware of not overpromising what can be delivered. As with anything that is made, the consequences of how factors such as environment, materials, design, and so on affect quality needs to be considered when designing supply chain models. On the supply side, there is little point in advocating a CLM or CMI model, for example, if the customer doesn't have the infrastructure and personnel to ensure the required level of QC and QA. For situations where quality is not particularly critical, or where there is a reasonable level of freedom, then perhaps those models are realistic. For more precise needs, while integrated fabrication options, such as Additive Industries' MetalFab1, are heading toward self-contained solutions, they are many years away from allowing anyone other than highly trained operators in highly controlled environments to produce finished items. On the demand side, it must be considered, for example, that while complexity is free-ish and mass customization is possible, there are consequences in lead times and costs involved in making things with the expected quality. Furthermore, including QC and QA processes adds to the time to manufacture items as well as their cost, potentially as much as 25% of the total. Those need to be included in any business case model.

⬛ Key Actions for Supply Chains

1. Identify what the quality requirements are for potentially 3D printed parts. This includes understanding the customer promise and recognizing any regulatory or other legal needs, such as health and safety controls.

2. Work with the 3D printer and materials providers to identify what QC and QA can be implemented both within 3D printing suites and the end-to-end processes.
3. Take the limitations and capabilities of 3D-printing QC and QA when deciding to use that set of techniques. This includes reviewing the terms of agreement with customers on quality.

STANDARDS

A significant hurdle in the pursuit of wider adoption of 3D printing, particularly in the areas of interoperability and QA, is the lack of a coherent, universal set of standards for 3D printing. As in other manufacturing areas, standards—the "agreed way of doing something," as the British Standards Institute describes them[12]—guide what is acceptable and what can be expected, particularly in technical terms. They provide a formalized and recognizable framework for QC and QA. For instance, standards provide a good measure of assurance that two sets of tools will produce roughly the same output when used in the same way. However, despite the technology being over 30 years old, 3D printing lacks an established and widespread set of standards. At present, the promise of consistency across two 3D printers operating in the same circumstances is left to the printer providers' promises and statements. As with the absence of strong QC and QA, this lack of global standards exposes the makers of 3D printed items to the risk that, for example, what they produce isn't what they thought they were making, or that things made on the same machines, whether in the same or different geographies, under similar or different circumstances, will not come out with the same characteristics.

The 3D printing and manufacturing communities have recognized this problem, and have made several attempts to set standards, although in a largely decentralized manner. In 2009, the US-led standards organization ASTM International formed Committee F42, a group of representatives from new and established businesses, trade associations, academia, and government whose mission was to develop standards for additive manufacturing.[13] Today, the output from F42 covers 3D printing processes and equipment, and 3D-printed finished parts, enabling certification and compliance with environmental, health, and safety legislation. In 2016, ASTM International partnered with the ISO's Technical Committee 261 (TC261), which had similarly been developing its own standards, to produce the Additive Manufacturing Standards Development Structure, a framework aimed at providing a unified set of technical standards in 3D printing.

That year also saw the launch of another initiative, the Additive Manufacturing Standardization Collaborative (AMSC), by the organizations America Makes and the American National Standards Initiative (ANSI). This cross-sector coordinating body was aimed at catalyzing the development of industry-wide additive manufacturing standards and specifications. Working with over 150 entities to ensure that all needs are addressed, from OEMs, government, and academic bodies to several standards organizations, AMSC has the ultimate goal of facilitating the growth of the additive manufacturing industry by developing and implementing 3D printing standards. In February 2017, the body published the first release of the "Standardization Roadmap for Additive Manufacturing" which identified existing standards and specifications, and those in development, identifying 89 gaps in all of those, and made recommendations for priority areas where there was a perceived need for urgent and additional standardization.[14] This included a wide range of topics in the 3D printing space, including:

- Design, including methods and tools
- Processes and materials, including process control and postproduction processes
- Qualification and certification programs
- Nondestructive evaluation methods
- Maintenance approaches

This was a first step, and refining the roadmap is an ongoing exercise. Other efforts have focused on particular areas and sectors. For example, 3D printing classification guidelines were published by the QA and risk management firm DNV GL in late 2017.[15] Their class guideline, DNVGL-CG-0197, offers companies in the maritime and oil and gas sectors a means of certifying parts before use, a critical issue for those safety-conscious industries. It covers the full lifecycle of 3D printing, starting with the materials employed in making parts, through the 3D printing technologies and processes used, to postproduction finishing. It also covers the transfer of data during those stages. This approach is also needed in other sectors, particularly the automotive, aerospace, and medical industries, which have launched their own certification frameworks. Naturally, the situation of having multiple standards bodies with different, if similar, standards presents its own set of challenges, much as it does in other industry sectors.

As well as standards for the 3D printing processes, there is a need for industrial standards for the materials used. These include codified standards for quality and quantities of materials, such as particulate size, purity, and a range of other physical and chemical characteristics. Currently, different materials suppliers offer powders and plastics in a variety of quantities and specifications, some of which tie specific materials to particular printers, which makes the task of procurement harder when it comes to qualifying the materials for sourcing. This need for materials standards has also been recognized by ASTM International, which embarked on issuing this guidance through its F42 group in 2018, specifically for metal powders, within the auspices of its wider standards set and the AMSC. In the medical arena, the US Food and Drug Administration (FDA) released its first official guidance on using 3D printed devices in its 2017 document, "Technical Considerations for Additive Manufactured Medical Devices." In that, the FDA articulated guidelines on the entire 3D printing ecosystem, from design through materials to production and postprocessing. It sets out the technical considerations when designing medical and dental devices for market uses, how those should be made, from what, and how they are to be finished. This includes advice on testing and ensuring that due quality is delivered.

One other area of 3D printing where standards are required is data management. Building on the question of format, discussed in Chapter 2, this must cover the definition, integrity, resilience, and security of data. In the short term, existing standards concerning data, such as the ISO 8000 (Data Quality and Enterprise Master Data), ISO 9001:2015 (Quality Management), and ISO 27001 (Information Security Management) sets, are typically applied to data by those working in the 3D printing sector.

Despite the number of initiatives seeming to imply the ease of setting these standards, the challenge is not insignificant. The pace of 3D printing technologies development, from data formats through 3D printing techniques, to the supply chain models that employ it and the needs of those, means that developing a comprehensive and definitive set of standards is a dynamic and evolving task. This doesn't mean that standards should be ignored, though: the frameworks and approaches that have been developed so far act as a known quantity to guide those who use 3D printing. Where there isn't a specific technical standard in place, companies have consulted with the standards organizations or built on existing approaches to give some measure of assurance. For those involved in supply chains, this means working with their wider organizations to develop standards that expedite that organization's goals in a unified manner. A good start is to identify the needs that are to be met, how they will be achieved, and what the constraints are. These then serve to inform what existing standards can be used and how to develop new ones, guided by methodologies such as the pioneering work proposed by Deloitte's Brian Tilton, Ed Dobner, and Jonathan Holdowsky.[16]

Key Actions for Supply Chains

1. Identify which standards are required for items made traditionally. These may stem from industry regulation and accreditation bodies, legislation, or safety agencies.
2. Review the latest standards sets from national and international bodies, such as ASTM International's F42, ISO TC261, the AMSC and others, to identify which standards best fit the particular needs of the supply chain.
3. Once a particular set of standards is agreed upon, ensure that a "tiger team" of informed and empowered representatives from the relevant parts of the supply chain is convened to delineate how, where, and when the standards will be implemented and to drive that adoption.

 ## REGULATION AND ACCREDITATION

In many sectors, the QA of 3D printing items is controlled through regulation, with processes and what those produce needing to be accredited and certified before they can be used as intended. This regulation can apply to entire items, most obviously to firearms, as well as their individual components. Before parts can be used on an aircraft, for example, they must be accredited for use by national and international aviation safety agencies to ensure and certify that they are safe, and that they were made with the required quality and to exact specifications. 3D printing presents a challenge to regulation and vice versa.

If viewed purely as a manufacturing technique, a 3D-printed part destined for an aircraft is no different from one fabricated via any other manufacturing approach, in that it must go through the same procedures for accreditation. If a manufacturing process is conducted at a site, independent of whether traditional or additive techniques are used, then that process and site must be regulated. Therefore, as a starter, accreditation procedures need to evolve to take 3D printing techniques into account, both for the parts and their supply chain, from the materials to the finished parts.

These efforts are underway. In the aerospace sector, for instance, accreditation bodies such as the Federal Aviation Administration (FAA) certified parts for Airbus and Boeing aircraft, and the lessons learned in doing so are being adapted in other sectors. The US Marine Corps is similarly working toward such accreditation for titanium and steel parts on the V-22 Osprey, H-1 Huey, and CH-53K rotorcraft platforms.[17] Similarly, devices produced for medical and dental purposes have to undergo stringent tests and confirm with strict standards regulated by national bodies such as the FDA in the USA and the National Institute for Health and Care Excellence (NICE) in the United Kingdom. Given that the healthcare sector is one of the biggest users of 3D printing, with the technology being used in ever more widespread situations, those bodies must react and adapt to cope with the new paradigm.

The bigger challenge, though, stems from controlling who can print things in the first place. Consider the scenario where an aviation manufacturer creates spare parts that have been certified using a 3D printer installed overseas. If they follow the exact steps that the original manufacturer employed, does each part require separate, on-the-spot certification, or is approval of the design and process employed sufficient? How is that enforcement carried out? These scenarios will be analyzed by those regulatory bodies, with consequential changes needed to legal and judicial systems. A similar quandary affects the concept of using 3D printing to dispense pharmaceutical products. One solution is to 3D-print medicines under the supervision of a licensed pharmacist who can monitor and control the end products more closely. The good news is that industry regulators are moving in the right direction: the FDA approved the first 3D-printed drug in 2015.

Key Actions for Supply Chains

1. Identify the current areas in the supply chain impacted by regulation and that require accreditation.
2. Review the regulation and accreditation schemes to highlight guidance on using 3D printing and the processes for demonstrably achieving necessary standards.
3. Work with 3D printer providers and the regulatory and accreditation bodies to select the right 3D printing solutions.

HEALTH AND SAFETY

As with any mechanical process, 3D printing presents a raft of health and safety hazards that can be divided into two groups:

1. Those related to the 3D printing process
2. Those related to what it produces.

Giving Flight to 3D Printing

The aerospace sector has long had an interest in 3D printing, primarily due to its ability to produce parts that are lighter and stronger than those made with traditional manufacturing techniques. The savings in fuel economy by reducing aircraft weight quickly add up and can reach hundreds of thousands of dollars annually if not more. If a reduction in weight of 1 kg saves US$ 2,500 per year in fuel costs for one aircraft, over a 30-year lifetime that is equivalent to US$ 75,000. If 1,000 parts on every aircraft were each made 150 g lighter, that would add up to fuel savings of US$ 11.25 million over the lifetime of that aircraft.

Despite the mathematics making a strong case, the restriction to using 3D printed parts long lay in the need for them to be accredited, much as any other part going on an aircraft requires. Existing accreditation schemes, such as those from the FAA and the European Aviation Safety Agency (EASA), were not written to take additive techniques into consideration, and neither were those techniques able to produce items that could be considered to have the necessary characteristics. However, with the advances in 3D printing techniques and further studies into those, the accreditation schemes and 3D printing technologies have begun to align. Consequently, both major aircraft manufacturers, Boeing and Airbus, have pressed ahead with 3D printing for their newest models.

Boeing has included several 3D printed parts in its 787 Dreamliner, the latest model to roll out of its aircraft production facilities, and Airbus is incorporating over 1,000 3D printed parts in its A350XWB. Both manufacturers received FAA and EASA clearance to 3D-print titanium structural parts, giving their airframe greater strength with lower weight than would otherwise be possible. Other manufacturers, such as engine providers GE and Pratt & Whitney, are also using 3D printing in the wider aviation supply chain. As costs and supply chain models evolve, manufacturers such as Embraer and Bombardier will likewise use 3D printing in the production of their aircraft and spare parts. New developments, such as the ability to 3D print in carbon composites, also promise to accelerate its adoption. Recognizing this, the 3D printer providers now sell equipment specifically aimed at the aviation sector by being FAA and EASA accredited.

Decision makers need to understand both when introducing the technology in their supply chains. This section will look at each in turn.

First, consider the 3D printing processes themselves. Unsurprisingly, these can be hazardous to health: many of the technologies involve materials at high temperatures, others employ lasers, and most require the use of high electrical power to supply them. As with any such engineering equipment, the right precautions need to be taken when handling materials and operating the machinery. Raw materials can be heavy, particularly metal powders, and the limitations—legal and practical—placed on human workers means that robots, forklifts, and other lifting gear are often needed.

In many cases, those raw materials are highly toxic. Thermoplastic and powder-based materials give rise to the nano-sized dust particles from raw materials, and sintering technologies to fumes. These nanoparticles are very small—less than 1/10,000 mm—and have large surface areas, which means that they can readily penetrate, interact with and pass through the body's systems, from skin and lungs, to nervous and brain tissues, at the cellular level. Those result in cardiopulmonary and respiratory issues, cancer, asthma, and nervous system effects. Early studies of the dust and vapors from 3D printing, such as the research conducted by the Illinois Institute of Technology and published in 2016, identified that they can present carcinogenic risks to those nearby.[18] Other studies are also looking at those risks in more detail, with pioneering work being carried out at the University of California Berkeley by a team led by Justin Bours and Brian Adzima.[19] Other investigations are studying the health risks from different types of materials, powders, and filaments. For instance, burning some plastics produces formaldehyde, a highly irritating gas that is also carcinogenic. As well as being toxic, many powders are liable to spontaneously ignite when exposed to air, and so present a fire hazard during handing. Those processes that use laser or electron beams usually use inert gases that displace oxygen from the objects being sintered, gases that are most often colorless, odorless, and harmful, and that incur a risk of suffocation in unventilated environments. These risks need to be considered when siting 3D printing equipment, particularly when placing it in a small room or confined space, or in areas where air quality needs to be strictly controlled, such as onboard submarines and on offshore drilling platforms. For instance, oxygen monitors may be needed, and access to the area controlled to minimize exposures.

Those involved in handling materials and items in a 3D printer environment clearly need to be appropriately prepared and dressed. This means receiving training in safety hazards, protocols, and processes, and using the right personal protective equipment (PPE), such as goggles, gloves, breathing masks, and even full airtight suits. The processes and procedures required for the end-to-end 3D printing process should cover:

- How to work with the materials and their containers
- How to access 3D printing equipment and its component parts
- How to dispose of unused and unnecessary material
- How do deal with emergencies, usually with the establishment of a set of emergency operating procedures (EOPs) for when accidents occur, such as material spillage or electrocution.

These should include, for example, ventilation procedures and minimum wait times before entering spaces or handling items, even though these will slow down the end-to-end production time.

The makers of 3D printers recognize the risks and are responding accordingly. Many printers now incorporate sensors to monitor particulates and gas concentrations. They also have built-in ventilation, carbon scrubbers, filters, extractor fans, and covers to protect those around the machines, reducing the risks when the machines are used in most situations, such as in manufacturing plants and schools. Other OEMs are going further and minimizing the need for human interaction by adding automation into the 3D printing equipment. This may take the form of robotic arms to pick up heavy containers of metal powders and load them into 3D printers, or to retrieve and separate items from build plates before moving them to subsequent treatment and postproduction stages.

These requirements for safety are often forgotten when discussing adopting 3D printing. Given the constraints they place on the production processes, supply chain decision makers need to consider them and actively address them. Ultimately, as with all manufacturing processes, there needs to be a safety culture that pervades the planning and implementation of 3D printing processes and that must consider aspects of production that may be new to many factory floors.

More widely, the environmental impacts of 3D printing, discussed earlier in Chapter 2, are only starting to be studied holistically. This is a point made by Reid Lifset, research scientist at Yale University School of Forestry & Environmental Studies:

> "Much of the research on the environmental dimensions of 3D printing has focused on benefits and impacts in production. That research needs to be complemented by more environmental analyses of the production of materials used in 3D printing and related upstream impacts, on how 3D products are used, and on the wastes (sic) they generate."[20]

Another area of health and safety concern lies in the emerging field of 3D printed organics, particularly those destined for consumption, such as food, and those aimed at medical uses, such a bio-printed organs and tissues. The first set introduce the need for a different type of safety certification as well as cleaning processes, much as would be the case with any other food preparation equipment. Bio-printing tissues, however, is a wholly new development, and medical qualification of the outputs will be subject to the types of restrictions that any other medical innovation has placed on it, including the standards and QA processes discussed earlier in this chapter. Moreover, the use of bio-printed organs—the goal of companies such as Organovo—brings with it ethical dilemmas, and the debate on whether to employ them has barely started outside a small group of specialists and bioengineers.

 ## Key Actions for Supply Chains

1. Consider the hazards of 3D printing when planning where to site and use equipment. This begins with including health and safety in the risk assessment at the outset of the planning process, and may result in 3D printers not being freely installed and used in some situations.
2. Work with operations and 3D printing providers to develop and implement the right risk mitigation actions, from the preparation of standard and EOPs to the provision of ventilation, air quality monitoring, and other physical controls.
3. Ensure that all personnel are suitably aware of the health and safety risks, trained in the correct procedures, and equipped appropriately to work with 3D printing equipment and materials.

 ## DATA MANAGEMENT AND SECURITY

The volumes of data that humanity creates today are astounding, and the pace is accelerating as the Internet becomes more widespread with more devices connected to it. We're all generating more data individually, from social media to office documents, from photographs to movies, and that is equally so for supply chains. As mentioned in Chapter 3, according to IBM, humanity created 2.5 quintillion (10^{18}) bytes of data in 2017 and 90% of all data ever created was produced in the previous two years.[21] Supply chain operations and the functions around those are all becoming ever more digital, and firms are now awash with data. Managing it is an enormous challenge, requiring investment in IT infrastructure, data processing, services, and high skills from even the smallest of enterprises.

However, not all data is digital. In many instances, key design data remains in physical written form, archived in drawings, long documents, and on data cards. While many firms have transitioned their operational processes to the digital sphere, such as procurement to payment, item designs are frequently not. This is particularly the case with long-lasting items and obsolete parts. A critical implication of 3D printing is the need to digitize that information, to convert those design drawings to CAD data files so that they can be used or modified to make new items. This requires the right investment, tools, and skills, and can be a lengthy process.

It also requires the right approach to data management. Those digital files quickly become unwieldy, and as 3D printing becomes more sophisticated, the size of data files will continue to expand. In parallel, data and the need to manage it emerge from the modeling, simulating, testing, and quality management processes that support 3D printing. The data volumes and processing speeds needed, such as for modeling the behavior of metal powders in a Selective Laser Sintering (SLS) system, mean that high performance systems are required to handle them, and those are scarce and expensive. If these sorts of simulations are to be carried out for every design, new approaches will be needed that expedite or simplify them.

Additionally, the datasets that are created during the fabrication processes themselves have to be taken into consideration. An increasing trend is for 3D printing systems to employ real-time monitoring, QC and QA, such as using video to ensure the correct application of laser beams in SLS machines, and these, too, quickly generate large volumes of data that also needs to be managed, requiring processing power and storage. Supply chain decision makers shouldn't be surprised if, having decided to use 3D printing, they are faced with needing to procure data management and processing capabilities in the form of equipment and skills, far beyond their existing ones.

With the need to manage data comes the ancillary consideration of its security. Today, the importance of data security is largely recognized when dealing with, for example, customer and personal data. Mismanagement of those data has serious consequences and many companies have seen damage to their reputation and share prices because of the mishandling of sensitive and valuable data. As demonstrated by the earlier example carried out by New York University, it is quite possible to sabotage designs and weaken items, potentially with catastrophic results. Processes, tools and behavior are all used to counter that threat. The introduction of 3D printing into a supply chain's operational model adds to the need for data security, particularly with regard to design data. Certainly, encryption of design data and being careful with sharing those are essential, much as personal data is protected using a mixture of training, process and tools. Controlling access to the relevant data storage devices is also essential, although more technical means may help, such as biometrics.

In most cases, it is the IP associated with designs that imparts the value of products, much like the famed secret recipe for Coca-Cola, and this makes digital designs of products far more valuable in their own right. Given that several supply chain models involve the transmission of digital design data beyond their boundaries, such as when using a CMI model or outsourcing the manufacture with a 3D printing hub, that data will need to be protected using encryption processes and tools, much as any other commercially confidential information would be. Moreover, where there is a threat of designs being modified for illicit reasons such as industrial sabotage, or stolen and sold to rivals, then controlling access to those designs is critical. These questions will naturally be of concern in those sectors that use 3D printers to make items for use in safety-critical situations, such as aerospace and defense, but equally to companies that could be prone to intense competition. Reflecting these needs, companies such as LEO Lane and Create It REAL have begun to offer 3D printing platforms that bring encryption between designers and printers.

This issue of security also extends to physical security. The case of the 3D-printed handgun and other weapons components has already indicated that security is a concern; consider also the ease with which components for an IED can be made using a 3D printer. The threat also extends to less violent criminal activity. In 2016, the security firm G4S announced that it had identified a first instance of thieves counterfeiting security devices to steal cargo, including ISO 17712 high-security seals, locks, and padlocks. In one case, thieves 3D-printed copies of security devices, which they then used to mask the theft of pharmaceutical goods from containers in a port.[22] In 2015, the CAD files for high-security Transport Security Administration (TSA) keys were published online, allowing anyone to 3D print a master key that opens any luggage lock. The international hacking community has also 3D printed keys to open police handcuffs, and developers at MIT CT-scanned complex locks as a precursor to printing keys to open them.[23] In response to these sorts of threats, innovation in the security field is taking 3D printing into account. One such enterprise, UrbanAlps, was founded by mechanical engineer Dr. Alejandro Ojeda to develop a new design for security keys—termed the Stealth Key—which conceals the key's teeth inside the structure (Figure 8.3).

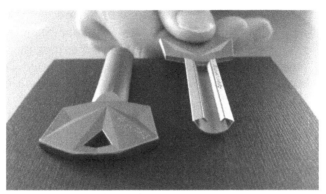

FIGURE 8.3   The 3D-printed Stealth Key.
Source: Reproduced by permission from UrbanAlps AG.

Not unsurprisingly, this complex design is only achievable by using a Selective Laser Melting (SLM) 3D printer to manufacture the keys.

As well as active security measures by companies, changes will be needed to prosecute data breaches and the mishandling of 3D printing data and items. This will include changes to legislation, as implied in the previous section, and to law enforcement—police forces need to understand what the threats from 3D printing are as well as the potential (mis)uses. The Abu Dhabi police force, for example, is ramping up the education of their officers on what 3D printing is and how to respond appropriately when it is used for nefarious purposes.

Key Actions for Supply Chains

1. Identify what the current state of data is across the supply chain. Consider whether to digitize designs that aren't so already. This can emerge from internal plans to adopt 3D printing as well as external needs from supplier and customers.
2. Work with IT departments to carry out data risk management. This must encompass how data will be stored, handled, altered, and protected, with clear policies, processes, tools, and training on those defined and implemented. Moreover, contingency plans to cope with data breaches and illicit interference should be prepared.
3. Consider the security implications of 3D-printing-enabled supply chain models, such as the need for data encryption when transmitting data to third parties, such as printer hubs.

4. Consider the needs for physical security in a 3D-printing-enabled supply chain, from control to equipment and data storage to the threats from 3D printed items that breach or bypass security measures.

COMMERCIAL MODELS

The music industry went through a change in the 1990s when the means to listen to music transformed from the physical (e.g. records, cassettes, CDs, minidiscs) to the digital. As MP3 files and entities such as Napster overtook sales of traditional media, music publishers were forced to rethink how to charge for music, an experience soon extended to movie distributors when the advent of broadband Internet speeds enabled the transfer of larger files. No longer were customers buying a hard, physical product: now they were buying digital strings of data, and companies that made and relied on the manufacturing and sales of physical media had to devise new ways of charging for that commodity. 3D printing requires the same revolution in commercial thought.

Companies and their supply chains are largely set up to charge for and transact in physical products, and those have one key advantage over digital things: the transactions and exchange of ownership are reasonably easy to monitor and control, with monies being exchanged from buyer to seller as that physical product changes owner. As happened in the music publishing world, manufacturing generally will have to revise the commercial models that it employs to charge for and control the use of its products when 3D printing is used, as that technology inevitably will be based on digital transactions, whether the supplier of that file is internal or external. Companies like Sculpteo and Shapeways already offer clients digital design files for 3D printing. As more suppliers move to 3D printing models, rather than selling physical objects, they will sell just the designs, which customers then produce in their owned, managed, or contracted 3D printing facilities. That move from physical to digital transactions means adopting a different set of processes and capabilities. For instance, the speed of the transaction is much faster, typically only constrained by the speed of Internet connections, and the systems needed to track the transaction need to be responsive to those speeds.

Key to that are considerations regarding the use of the data. Currently, when Ford sends a part to a service station, it can be used in one vehicle, and the price of the part is then set for that single use. A digital design file can just as easily be used to make one or one thousand parts. Therefore, part of the contract covering the transaction of 3D printing files will cover the license for their use. Typically, those can be:

1. Perpetual, allowing a buyer unlimited use of the software or data for eternity
2. Subscription, which only allows the buyer to use the software or data for a limited time or in a limited manner, such as to produce a limited number of parts
3. Consumptive, through which a buyer pays for the software or data according to how often those are used

Building on the legal, data management, and health and safety issues described earlier, the licenses can be passive, using an honor-based agreement, or more active, switching off access to data or its use, or even going so far as to erase it entirely. As many in the music, movie, and software sectors have done, supply chains will need to adapt their supporting functions to manage those licenses, requiring new skills and systems.

The needs of this new paradigm are driving users to innovative solutions, such as the use of the much heralded blockchain, which seeks to provide the necessary trust, assurance, speed, and transparency needed to carry out digital transactions. Blockchain also has the potential to provide traceability, integrity, and assurance throughout a supply chain that incorporates 3D printing, particularly with regard to controlling the use of design files and their distribution. Moreover, the mechanisms by which it can do so can be used to provide a basis for the required new commercial models. That said, blockchain is a still a very immature technology and a good deal of caution is needed before embracing it.

Key Actions for Supply Chains

1. Define how 3D printing will be used in transactions. For example, will designs be created and fabricated in-house or sold on to customers?
2. Review and adjust the commercial policies, processes, and tools to handle the needs of a data-centric transactional basis. This includes considering how to control the release of digital designs and manage their use.
3. Design and implement a licensing approach that defines precisely how design data can be used. This much include the definition of legitimate use and allowance for production of items, as well as the sanctions that will be imparted if the agreement is broken.

4. Consider the use of innovative solutions to manage digital transactions, such as the use of blockchain-based approaches.

FISCAL AND FINANCIAL IMPLICATIONS

3D printing is not just an issue for companies; its introduction will also affect governments, and its implications will come to be felt in supply chains. New digital models, such as those espoused by firms like Airbnb and Uber, are challenging tax authorities on how to deal with them, and 3D printing is widely seen as the most difficult of these issues to resolve. For instance, the simplification of supply chains, with 3D printing reducing the need for assembly and thus for different suppliers and steps in the process, means that they will incur a lower tax burden and fewer intercompany transactions. When trading across borders, manufactured goods incur several taxes, particularly tariffs, often to protect national industries as well as to raise tax receipts. What happens when 3D printing enables overseas manufacture without physical items moving across international boundaries?

For example, today Brazil levies a 35% import tax on top of a 55% industrial tax on US-made auto parts coming into the country as a way of protecting local industry. If companies such as Ford and GM succeed in their aspirations of 3D-printing those parts, allowing them to transmit digital designs to locally placed printers, then those revenues evaporate and all the barriers that exist to protect Brazil's auto parts industry are demolished. For Ford and GM, though, this situation presents them with an attractive proposition: use 3D printing to manufacture parts to their designs in tariff-unfriendly countries, and the cost of those parts drops, enabling them to price those parts far more competitively. The question of whether transmitting a digital design from a supplier to a customer represents a sale or a licensed transaction will affect its treatment for withholding tax, for instance. Certainly, companies will repeat the behaviors of the past and set up their operations with arrangements that are tax efficient. They will employ different tax treatments in different geographies, such as establishing entities in tariff-free zones from which they simply transmit digital designs, thereby minimizing—if not eliminating—tax bills.

Experience shows that government policy and fiscal legislation often lag technological advances, and policy makers frequently get overzealous in tackling the gap. Consider the adoption of the Internet in Brazil, Russia, and China, where policy makers have, at several points, attempted to bring controls to Internet use, from shutting down access to applications like WhatsApp or websites such as LinkedIn, to legislating against free speech on social media. It is quite possible that, when faced with diminishing tax revenues and increased competition, policy makers go after what they can, such as placing taxes on 3D printers and printing materials, as well as on each printed item itself. Sooner or later, local and regional governments will find ways to deal with threats to tax and customs revenues. Ernst & Young tax partner Channing Flynn highlighted this in 2016:

> "Each of the potential business benefits of 3D printing carries tax implications that could alter the equation for any anticipated operating efficiency or return on investment."[24]

He went on to add:

> "The tax risk companies face is already at an all-time high worldwide, with global digital business models posing unprecedented challenges to tax authorities and provoking conflicting tax policy from country to country."

Supply chains considering 3D printing items are therefore well advised to monitor government actions and include changes to the fiscal environment on their risk register.

Despite the uncertainty of the longer-term tax picture, several governments are looking to grow their national 3D printing capabilities, often by offering tax incentives to support research and development (R&D). These promotions have been already leveraged by firms using 3D printing technologies during the prototyping phases of product development. As supply chains move 3D printing from R&D through piloting to full-scale production, engagement with tax departments can make that journey less costly and more tax efficient. The result of those discussions will be a factor shaping the supply chain, much as it does today, addressing issues of where to locate 3D printing and how the transfer of IP between financial jurisdictions is treated and incentivized.

▨ Key Actions for Supply Chains

1. Work with tax departments to map the tax implications—beneficial and costly—of using 3D printing in the supply chain, and maintain a close watch on emerging changes from national, regional, and local governments.

2. Consider tax incentives when shaping the 3D-printing-enabled supply chain.
3. Be clear on how design data IP is transferred between entities within a supply chain and with those outside of it, such as whether it is a sale or licensed transaction.

SKILLS BASE

When the 3D printing firm Sculpteo released the findings of their 2018 survey of the state of 3D printing,[25] it revealed that the biggest change that companies thought they should make to fully capitalize on 3D printing was to "increase expertise and education of collaborators." Much as the use of robotics has changed the face of manufacturing, 3D printing brings with it a shift in the types of roles that companies will need to fill. As manufacturing migrates to 3D printing, so the use of highly skilled technicians to operate lathes, presses, and other traditional techniques will decline. There will also be concurrent rising demand for highly skilled CAD designers, 3D printer operators and maintainers, and IT management resources. Even where some of those exist today, such as CAD designers, they need new skills and training. Designing for 3D printing calls for different approaches, unconstrained by the limitation and considerations of traditional manufacture, something that has been termed Design for Additive Manufacture (DfAM). Whereas before they were concerned with how to work blocks of materials to achieve the dimensions they wanted, now they can create things more freely.

Of course, there are other design considerations, other rules, to take into account if manufacturers want to get the most from the 3D printers during production. For instance, they must learn to design products that transition gradually between adjoining surfaces to avoid weaknesses. They will need to minimize large differences in cross sections and part volumes. Sharp corners that cause residual stress, bucking the finished items, need to be reconsidered. Similarly, the inclusion of unsupported thin walls weakens the structural integrity of objects. With many 3D printing technologies, having shallow surfaces results in stair-stepping: surfaces will have a rough texture as the printer heads moves along to make them. Designers also need to have a good technical understanding of the 3D printing technologies that will be used to make their designs and how they can be finished in postproduction.

The number of people required for a 3D printing-based production line is much smaller than for a traditional assembly line. With 3D printing reducing the total separate elements in an item, simplifying the steps involved to produce it, and increasing the amount of automation that 3D printing processes employ, the staff involved in making things is significantly reduced. It also reduces the demand for personnel in supply chains overall, as many workers involved in executing tasks in logistics, from warehouse staff to transportation, will no longer be needed in the same numbers, and several will be reskilled to support 3D printing instead. When you combine these two impacts, the numbers are staggering: 10 to 20 jobs in a traditional assembly line are replaced by one.

While undoubtedly an opportunity to reduce personnel costs, this shift also presents a challenge to supply chains looking to adopt 3D printing. The demand for skilled labor is already exploding: according to John Hornick, the partner at law firm Finnegan and a 3D printing commentator:

> "Skilled 3D printing–related jobs soared 1,384 percent from 2010 to 2014 and were up 103 percent from 2013 to 2014."[26]

That trend has continued as 3D printing expands into new supply chains. Hornick adds that the most called-for categories are industrial and mechanical engineers, and software developers, and this is supported by subsequent surveys such as that conducted by Sculpteo. Companies will have to change what they look for in those employees, as well as where they find them. As with other emerging industrial technologies, rather than being unskilled, those employees will increasingly be university graduates and skilled apprentices, which means that pay scales will be higher. Moreover, the talent pool from which to recruit is not big, and other parts of the manufacturing industry will also be seeking personnel from it, further driving up staff costs. In most of the leading manufacturing countries, there is already shortage of skilled manufacturing labor, and responding to that is a matter of urgency, one that many firms recognize. A 2017 survey by the consultancy RSM identified that skills shortage was a major challenge for 43% of UK manufacturers,[27] while the consultancy Deloitte calculated that, during the period 2015 to 2025, three and a half million manufacturing jobs will need to be filled but that two million of those will not be, a shortfall of 57%.[28] This is exacerbated by the increasing average age of those employed in manufacturing jobs worldwide, and a general insufficiency of STEM (science, technology, engineering, and mathematics) education and training, particularly in Western economies.

Beyond the manufacturing skills, software skills will be highly sought, and software engineers will be at a premium. In an interview with the Innovation Centre of Western Australia, John Hornick said:

> "They will be in high demand to write, update, and manage software to meet 3D printing-related software demands."[29]

He includes in those the need for solutions "for anti-counterfeiting; authenticating parts and products; authorizing, certifying, controlling, and managing service providers; converting designs and scans to 3D-printable data; customization; design; digital rights management (DRM); encryption; file authentication; [and] file management." Of course, there will still be a call for the sorts of skills that software-enabled technology always calls for to operate and control high-tech machinery. Those software skills, however, also will be required by supply chains undergoing broader digital transformation and implementing other technologies, from artificial intelligence and blockchain, to supply chain "control towers" and data analytics.

To meet that shift in demand from industry, education and training will have to adapt if national workforces are to supply it. This means offering courses and modules that develop the academic and practical skills needed. It means driving to increase the numbers and level of education of those pursuing courses in STEM subjects. Moreover, there needs to be a shift in perception of manufacturing from one where it is seen as manual, dirty, and repetitive. Here, the benefits of 3D printing may play an important part if it is positioned as being more focused on design, high tech and more engaging. Until that happens, the adoption of 3D printing as a manufacturing process is likely to be constrained more by a lack of suitable, qualified, and experienced people than by the restrictions of the technology.

As well as nurturing the skills needed for 3D printing, companies will have to respond in the shorter term, re-skilling and retraining their existing workforces that are deeply vested in traditional manufacture, as well as growing their own talent through apprenticeships and other similar programs. With the flexibility in design removing normal constraints, with new materials presenting different properties, and with the new dynamics of a 3D-printed manufacture approach, those long-experienced designers and manufacturing staff will now have to adapt to and live in a 3D printing ecosystem. This is not something that happens quickly or cheaply, and investment is already needed to ensure that the staff concerned are ready for what is coming. One course of action to consider is to pilot 3D printing, using a 3D printer hub, for personnel to gain an appreciation of what it can do. This will also help in identifying internal skills gaps more precisely.

The shift in skills and the lower number of people involved clearly have societal repercussions. While today some schools and colleges, as well as university courses, introduce students to 3D printing, there remains a lacuna of understanding of what it is, how important it will be, and what it will require in terms of skills. This needs to change. Academic courses must include CAD/CAM/CAE and 3D printing in at least the secondary level of education, and engineering and design courses need to be adapted to put the technique in their curricula. Those changes will have to be financed, from changing syllabuses to training teachers and investing in 3D printing equipment in schools (and thereafter maintaining them), and that money will have to come from government education departments or through industrial sponsorship. The D&T Association, for example, managed a UK Department for Education–funded program on 3D printing with 42 teaching schools, covering primary, secondary, and special education institutions. In that program, each school was given a 3D printer, although the schools themselves had to develop teaching materials and activities to help other schools to leverage the technology. If companies want to ensure that the workforce of tomorrow is ready for 3D printing, then efforts to sponsor schools to have 3D printers and to teach the pupils about the technology and what it can do will have to grow. Such efforts will include apprenticeship programs, such as those launched by the Manufacturing Technology Centre in Warwick, UK, which also holds the country's National Centre for Additive Manufacturing (NCAM). That program counts on the combined support of dozens of industrial companies which will provide industrial placements for those apprentices, giving them the practical experience needed to train tomorrows specialists.[30] This follows recent investments from the Robert C. Byrd Institute for Advanced Flexible Manufacturing, which funneled US$ 4.9 million to expand its model apprenticeship program across the USA, with particular emphasis on the training of additive manufacturing.[31]

▦ Key Actions for Supply Chains

1. Identify the current manufacturing skills base in the company and identify what 3D-printing-related skills already exist. This includes the technical skills in the wider supply chain.

2. Work with 3D printer providers and operations personnel to map the skill needs in the end-to-end supply chain that 3D printing requires, from design through production to postproduction and logistics.
3. Consider re-skilling and retraining existing personnel who will be required in a 3D-printing-enabled supply chain.
4. Establish personnel career management plans, identifying where talent will be obtained, and consider collaborating with educational and academic establishments to develop courses.

The restrictions of 3D printing and the wider implications of its adoption are being steadily overcome, making it an increasingly attractive option with the potential to be a strategic differentiator. As the technology evolves and matures, so its adoption in mainstream manufacture is accelerating. That leads to the question: how do companies go about considering whether 3D printing is the right solution for them and, if it is, how to they go about adopting it? That question is addressed in Chapter 9.

 ## Notes

1. Scott J. Grunewald, "When eBay Sellers Try to Defend Their Illegal Sale of 3D Models from Thingiverse, Comedy Ensues," 3dprint.com, February 20, 2016, https://3dprint.com/120727/ebay-licensing-3d-models.
2. Tala El Issa, "Jordan and 3D Printers: A Battle Close to an End," wamda .com, March 6, 2017, https://www.wamda.com/memakersge/2017/03/jordan-3d-printers.
3. Gartner, "Gartner Reveals Top Predictions for IT Organizations and Users for 2014 and Beyond," press release, October 8. 2013, https://www .gartner.com/newsroom/id/2603215.
4. Interview with the author, August 2017.
5. UK Government Office for Science, "Forensic Science and Beyond: Authenticity, Provenance and Assurance: Evidence and Case Studies," 2015, https://www.gov.uk/government/uploads/system/uploads/attachment_data/file/506461/gs-15-37a-forensic-science-beyond-report.pdf.
6. James Beck, "Unfortunately Disappointing 3D Printing Law Review Article," *Drug & Device Law* (blog), June 30, 2015, https://www.druganddevice lawblog.com/2015/06/unfortunately-disappointing-3d-printing.html.
7. International Standards Organization, ISO 9001:2015 Quality Management Systems, 2015.

8. Nick Hall, "Is 3D Printing an Open Invitation for Industrial Sabotage?" 3dprintingindustry.com, July 14, 2016, https://3dprintingindustry.com/news/3d-printing-open-invitation-industrial-sabotage-87150.

9. Sigma Labs, accessed December 10, 2017, www.sigmalabsinc.com/products.

10. Billy Short, "Quality Metal Additive Manufacturing (QUALITY MADE) Enabling Capability," Office of Naval Research, July 28, 2015, http://slideplayer.com/slide/6642475/.

11. Ian Wing, Rob Gorham, Brenna Sniderman, "3D Opportunity for Quality Assurance and Parts Qualification," Deloitte University Press, November 18, 2015, https://www2.deloitte.com/insights/us/en/focus/3d-opportunity/3d-printing-quality-assurance-in-manufacturing.html.

12. BSI Group. "Information about Standards," accessed July 31, 2017, https://www.bsigroup.com/en-US/Standards/Information-about-standards.

13. ASTM International, accessed September 25, 2018, www.astm.org.

14. America Makes & ANSI Additive Manufacturing Standardization Collaborative, "Standardization Roadmap for Additive Manufacturing," ANSI Standards Activities, June 2018, https://www.ansi.org/standards_activities/standards_boards_panels/amsc/.

15. DNV GL, "Additive Manufacturing—Qualification and Certification Process for Materials and Components," DNVGL-CG-0197, November 2017, https://rules.dnvgl.com/docs/pdf/DNVGL/CG/2017-11/DNVGL-CG-0197.pdf.

16. Brian Tilton, Ed Dobner, and Jonathan Holdowsky, "3D Opportunity for Standards: Additive Manufacturing Measures Up," Deloitte Insights, November 9, 2017, https://www2.deloitte.com/insights/us/en/focus/3d-opportunity/additive-manufacturing-standards-for-3d-printed-products.html.

17. John McHale, "3D-Printed, Safety Critical Parts Fly on V-22 Osprey," Military Embedded Systems, August 1, 2016, http://mil-embedded.com/news/3d-printed-safety-critical-parts-fly-on-v-22-osprey.

18. Parham Azimi et al., "Emissions of Ultrafine Particles and Volatile Organic Compounds from Commercially Available Desktop Three-Dimensional Printers with Multiple Filaments," *Environmental Science & Technology*, 50, no. 3 (2016): 1260–68, doi: 10.1021/acs.est.5b04983.

19. Justin Bours et al., "Addressing Hazardous Implications of Additive Manufacturing," *Journal of Industrial Ecology*, 21, no. S1 (2017), https://doi.org/10.1111/jiec.12587.

20. Reid Lifset, "Editorial. 3D Printing and Industrial Ecology," *Journal of Industrial Ecology*, 21, no. S1 (2017), https://doi.org/10.1111/jiec.12669.

21. Watson Customer Engagement, "10 Key Marketing Trends for 2017 and Ideas for Exceeding Customer Expectations," IBM Marketing Cloud, July 18, 2017, https://www-01.ibm.com/common/ssi/cgi-bin/ssialias?htmlfid=WRL12345USEN.

22. Hanna Watkin, "G4S Warning: Cargo Thieves Using 3D Printing," All3dp.com. https://all3dp.com/g4s-warning-cargo-thieves (accessed 30 May 2016).

23. Andy Greenberg, "MIT Students Release Program To 3D-Print High Security Keys," *Forbes*, August 3, 2013, https://www.forbes.com/sites/andygreenberg/2013/08/03/mit-students-release-program-to-3d-print-high-security-keys/#34d3cca973dd.

24. Channing Flynn, "The Questions Executives Should Ask about 3D Printing," *Harvard Business Review*, April 19, 2016, https://hbr.org/2016/04/the-questions-executives-should-ask-about-3d-printing.

25. Sculpteo, "The State of 3D Printing 2018," accessed September 25, 2018, https://www.sculpteo.com/en/get/report/state_of_3D_printing_2018/.

26. John Hornick, "3D Printing New Jobs," LinkedIn, April 16, 2016, https://www.linkedin.com/pulse/3d-printing-new-jobs-john-hornick.

27. RSM, "UK Manufacturing Monitor 2017," 2017, https://www.rsmuk.com/-/media/files/manufacturing/uk-manufacturing-monitor-2017.pdf.

28. Craig Giffi, "The Skills Gap in U.S. Manufacturing: 2015 and Beyond," Deloitte Press Report, 2015, https://www2.deloitte.com/us/en/pages/manufacturing/articles/boiling-point-the-skills-gap-in-us-manufacturing.html.

29. ICWA, "'3D Printing Is an Entrepreneur's Dream'—International Additive Technologies Expert John Hornick in an Exclusive Interview with the ICWA," accessed June 30, 2017, http://innovationcentre.gcio.wa.gov.au/3d-printing-is-an-entrepreneurs-dream-exclusive-interview-with-for-the-icwa-international-additive-technologies-expert-john-hornick/.

30. Manufacturing Technology Centre, "MTC to Launch Additive Manufacturing Apprenticeships," March 26, 2018, http://www.the-mtc.org/news-items/mtc-to-launch-additive-manufacturing-apprenticeships.

31. Charlotte Weber, "Apprenticeship Program Expanding Nationwide," *Manufacturing Engineering*, November 1, 2015, 110–11, http://www.sme.org/uploadedFiles/Publications/ME_Magazine/2015/November/November%202015%20Workforce.pdf.

CHAPTER NINE

Adopting 3D Printing

B Y NOW IT should be clear that 3D printing is not a passing fad or merely a technology for tomorrow: the revolution in manufacture is here, knocking on the door. In many sectors, it is already through that door, an integral part of supply chains. All indications are that this trend is accelerating: 1,768 metal 3D printing systems were sold globally in 2017 compared to 983 in 2016, an increase of some 80%.[1] For those companies already using it, the results have been impressive, if not outright transformative, and many industries are being radically changed today while others soon will be. As was the case with those traditional hearing aid companies that refused to adapt to 3D printing, firms that ignore it will simply be put out of business, overtaken by those that either prepare for it or use 3D printing to make better things, designed more closely to their customers' precise needs, with faster end-to-end cycle times, closer to those customers and at a lower overall cost. 3D printing will also play a growing role in mass manufacturing, lowering the cost and timescales in that area through its use in the manufacture of molds, while offering those firms using traditional manufacturing techniques a means to have jigs, fixtures, and other tools customized to the precise needs of their equipment both faster and more cheaply.

As with all major operational changes, the first step is to establish that there is a need for that change, that there is some threat or some opportunity, and then to act upon it. 3D printing is no different. In the introduction of this book, it was noted that the supply chain decision makers in many companies are asking questions of what 3D printing is, what it can do for them, and what needs to be considered. However, the wider firm may not be so aware of the situation and thus need to be convinced. Part of that comes from recognizing the impact 3D printing has on the environment around the supply chain, from suppliers to customer, as well as from the competition. With sufficient impetus behind it, the questions turn to how to take the first step into employing 3D printing in a supply chain. This chapter offers approaches for undertaking those.

After concluding that there is a case for adopting it in their supply chains, from a pilot project to a more wholesale roll-out, how to do so becomes a matter of operationalizing it. No two companies, and no two supply chains, are alike, however, even within similar industries. Trying to develop a single, detailed formula that companies can use to decide on whether to adopt 3D printing is, therefore, an exercise in futility. That notwithstanding, there are a series of consistent steps that companies can take to examine the questions of how, where, and when to adopt 3D printing in their supply chain. Those steps can be arranged in a framework to guide companies on the journey—the Supercharg3d Blueprint. This chapter describes what that framework is, guiding businesses on how to tackle the question of 3D printing in their supply chain.

 ## CONVINCING THE DOUBTERS

Today's supply chain decision makers are saturated with news about transformational techniques and tools that all promise to resolve one or several of the everyday challenges that they face. Blockchain promises to accelerate supply chains and bring end-to-end visibility. Control towers act to coordinate the materials and data flows to smooth out demand and balance supplies. Artificial intelligence cuts through the mountains of Big Data to provide predictive analysis of demand, optimize procurement, and increase service levels. Industry 4.0 builds on the digital transformation of manufacturing and connects supply chains. In that environment, those decision makers continue to face the age-old questions of how to get what customers want, where they want it, at the right time, and at the right cost. The challenge for 3D printing is how to make its voice heard in that cacophony of innovations.

It certainly doesn't help that 3D printing has been a much-hyped technology, despite being over 30 years old. Many supply chain directors and managers will have read the promises made in the last 10 years without witnessing the creeping conversion of promise into reality. With a basis in that coverage, they will still have visions of 3D printing that would befit science fiction stories and thus be highly suspicious about the reality. Others will have missed that 3D printing impacts the end-to-end architecture and processes of supply chains, thinking that it is merely another tool to make things in some "cool" manner. Still others may not be aware of 3D printing at all.

Of course, there will be resistance from some quarters, as has been the case with every major technological shift, from the Luddite protesters opposed to the advent of machinery in the textile mills of England in the 1800s. When automated teller machines first entered the retail banking sector, they were opposed and belittled by bank tellers, who later would come to be grateful for them as they removed the lesser value-adding, more menial tasks that previously had to be done by hand. For 3D printing, resistance is most likely to come from those who will see their skills at risk from the new technology, from the craftspeople and technicians involved in traditional manufacturing to warehouse staff. It will come from the middle management layers in design, manufacture, and the broader supply chain who will resist the redesign of their processes, feeling that those changes weaken their role and importance in those. Overcoming those objections requires time, communication, and involving the people concerned in the discussions in an open and transparent manner. The first goal is to convince sufficient decision makers and those affected to give the effort sufficient time and resources to be able to take the first step.

Of course, as this book has described, adopting 3D printing can easily be an organization-wide, strategic effort, something altogether daunting, and identifying where to start—what that practical first step is—is all too often seen as an impossible task. The key is to take an engineering approach, to start small and build on the lessons from that. The first step toward adopting 3D printing comes in making supply chain and operations decision makers, from CEOs to manufacturing managers, aware that it truly is a disruptive technology, that there is a threat to their continued performance, and that 3D printing offers potentially significant gains. John Kotter, the Konosuke Matsushita Professor of Leadership, Emeritus, at Harvard Business School, and a leading expert in what it takes to drive change in organizations, calls this "establishing a sense of urgency."[2] Gaining this recognition is essential for the continued success of efforts such as adopting 3D printing; without it, there will never be the engagement needed from those whose involvement is more needed, dooming the efforts to failure.

THE DISRUPTIVE RISK OF 3D PRINTING

The simple fact that the annual growth of 3D printing has been steady at 20 to 30% for over two decades means that it is radically changing the competitive landscape in many sectors and that this trend will accelerate. At the very least, businesses must consider it as a risk and assess its impact on them. Many firms clearly likely to be affected, such as the logistics firm UPS, have spotted the risks to their operating models and have responded by pivoting their business to embrace 3D printing within their service offerings. The disruptive risks of 3D printing can affect a business's entire ecosystem, and once it has identified and recognized this, it must establish the right mitigation and contingency plans. Those will range from dismissing the risk altogether, perhaps for one or two planning cycles while the technology matures, to actively adopting 3D printing as soon as is practicable.

This risk analysis should address all the external forces that affect a business, remembering that risks stem from things that result in change, both good and bad—all too often, risk assessments only consider what will negatively affect operations and business fortunes, ignoring the potential opportunities (Figure 9.1).

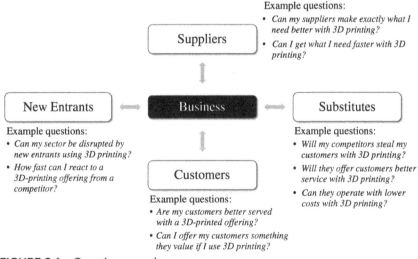

FIGURE 9.1 Questions to ask.

Customers

At the heart of every supply chain is the question of what the customer wants and how best to meet that. Independent of whether that customer is internal or external to a company, those wants will be expressed in some combination of functionality, cost, and time. To better meet those is a constant pursuit, often dictating the need for supply chains to adapt, using new approaches and technologies to do so. Increasingly, customers want a specific item for their specific needs, from a bespoke jig for a manufacturing line to a complex design of houseware. In some cases, the dynamics of a sector dictate what those customers look for, such as the pursuit for strength and lightness in aviation and space parts driven by the high costs of fuel. As 3D printing becomes more widely known, as its benefits are increasingly recognized while its constraints are solved or designed away, customers will begin to seek out the sorts of solutions that only 3D printing can provide.

To deal with the risk of 3D printing in terms of its customers, companies need to ask themselves how they could better meet their customers' needs by using the technology. Perhaps it is the opportunity to design products to their personal needs, such as an individually crafted cell phone case or a pair of bespoke and fitted sunglasses. Perhaps it is by offering the sort of advantages in simplification, fuel efficiency, and weight that drove GE to redesign the approach to making its Leading Edge Aviation Propulsion (LEAP) engine fuel nozzles. The approach to map that risk is to ask questions like, "Are my customers better served with a 3D-printed offering?" and "Can I offer my customers something they value if I use 3D printing?" This requires starting from a blank sheet, unconstrained by the limitations of how the company is tackling those today.

Outside of that, there are risks from customers pursuing 3D printing themselves in areas and with products that the company produces. In those instances, they may well ask their suppliers for more information on designs, including digital design data, and provisions for that must be made, from the digitization of designs to embedding the right provisions in contracts and licensing agreements. Moreover, they may want to be more involved in the design phases, looking for ways to simplify or perfect designs and reduce their own supply chain complexity. To deal with that, supply chains need to do what they should be doing anyway: engage their customers on the topic. This way not only can supply chains have visibility of the customer's direction, but new opportunities can emerge given the potential new paradigm that 3D printing offers.

New Entrants

The flexibility of supply chain models to ramp up the use of 3D printing to make things, such as by using 3D printer hubs, allows new businesses to gain a foothold far faster than has previously been possible. With 3D printing, the barriers to market entry that have previously curtailed the entry of new firms in so many sectors are quickly demolished. Rather than finding and investing large amounts of time and money, businesses can use the manufacturing capabilities of those hubs to establish themselves and open for business in a matter of days, offering products that can be customized to the precise, personalized needs of their customers. Those can be aimed at the consumer markets or specialist industrial products. In sectors where even the slightest advantage in customer service, speed, or cost can make the difference between growth and failure, the benefits that 3D printing brings, together with the speed with which a new entrant can be set up, means that start-ups can quickly change a market's dynamics. Of course, not every start-up will be the unicorn that corners the market in record time. However, the pace at which new entrants that use 3D printing are emerging, and their numbers, mean that there is plenty of opportunity for lessons to be learned and a successful model to evolve out of that landscape.

How to prepare for this? The first step is for companies to understand their market in detail, recognizing what the customers want and why they choose to buy from one company over another. Next, companies need to understand what it takes to enter the market and what the barriers are. This will enable them to grasp how likely it is for a disruptive new entrant to pitch their tent. Sectors such as aviation and defense, which have mountains of regulations to adhere to, are less likely to see such a start-up—it takes a long time to obtain air worthiness safety certification. Other firms, in sectors where products are complex in design, and having an offering that can deliver those quickly and cost-effectively makes all the difference, are more likely to see new 3D-printing-based service lines that quickly increase their market share. That is particularly the case in the healthcare sector (despite the need for regulatory approvals). For instance, the success of 3D printing in orthodontic firms such as Align Technologies (see sidebar) has driven new entrants such as SmileDirectClub, changing the dynamics of that sector.

Suppliers

Asking how 3D printing affects the supplier base inverts the questions used to analyze the disruptive risk from the customer perspective. Central to that is asking, "What can my suppliers do for me beyond today's capabilities if they switch to 3D printing or add it to their mix?" and "What are the impacts to me if they do?." Just as new entrants may threaten a company's position in the market, they can also provide an opportunity to diversify a supplier base. Moreover, having a supplier embarking on a path to use 3D printing opens the possibility of a better offering, particularly through collaborative development. Imagine if your supplier could deliver precisely what you need in a manner that is better for you.

As for all changes in manufacturing approach, it should be obvious that one consideration for suppliers using 3D printing is the need for them to provide quality assurance (QA) for what they make. Will the products have the same strengths, dimensions, and finish? What are the consequences on support for those parts? If suppliers reduce the number of components in an item, what is the effect of the change on the nature of the lowest replaceable part? If suppliers switch to 3D printing, understanding how they will provide QA and what is possible to QA with that technology become significant needs.

Substitutes

The biggest threat is likely to come from competing companies offering products similar to yours using the benefits of 3D printing. By leveraging advantages in cost, design, and time, competitors can gain an edge that traditional manufacturing cannot deliver. In this respect, GE led the way in the engines sector, changing the performance of its LEAP engines in a widely publicized way that catalyzed its rivals in Pratt & Whitney and Rolls-Royce to similarly seek improvements using 3D printing.

The main questions to ask, then, include:

- "How will my competitors steal my customers with 3D printing?"
- "Will they offer customers better service with 3D printing?"
- "Can they operate with lower costs with 3D printing?"

Having completed that external assessment, the next step is to consider internal factors. That means supply chains looking at their own inventory, to examine their own tools and components, and review their own design approaches and manufacturing methodologies, to see whether and where 3D printing can be used to enhance that particular part of the supply chain and where to reduce their overall cost. Most importantly, supply chains need to step back and consider whether 3D printing offers a way to change their operating model to gain the advantages that are called for.

Changing Smiles with 3D Printing—Align Technology

Align Technology is a medical and dental device company headquartered in California that in 2017 achieved a revenue of US$ 1.5 billion. Founded in 1997 by two graduates from Stanford University's MBA program, Zia Chishti and Kelsey Wirth (neither of whom was an orthodontist or had any medical training), the company chose to pursue a vision of digitized dentistry, enabled through a new approach to providing solutions for the realignment of patients' teeth.[3] Gone was the use of painful and unsightly metal shoes and wires, to be replaced by Invisalign, a clear-plastic-based system that received US Food and Drug Administration approval in 1998 and was first marketed in 2000. Although the materials and production approach would change over the first years of the company's operation, it has found success by investing in 3D printing technology to make the aligners that its patients need.

Traditionally, such plastic aligners use casts of a patient's teeth that are then manually adjusted by specialist orthodontists to produce a cast for a plastic aligner, with a new cast called for once or twice a month. Using thermoforming machines, each cast is then manipulated to produce an aligner, a process with several steps that must be repeated up to 48 times to create each iteration of aligner in a particular course of treatment.[4] This increases the cost and time of production, with patients having to visit the orthodontist's office several times. 3D printing changed that approach.

The new methodology begins with 3D scanners taking an impression of a patient's teeth, moving away from physical putties. The images from those are then worked on to develop the series of designs for successive aligners, each adjusting the patient's teeth just a little. Those designs are then sent to the firm's dedicated production facility in Juarez, Mexico, to be fabricated in one of over 50 stereolithography (SLA) 3D printers in the plant. This revamped process enables the company to produce over eight million aligners annually, or 220,000 per day;[5] by early 2018, Align Technology had helped over five million patients worldwide. By lowering the cost of manufacture, the firm can offer Invisalign for a price that is competitive against traditional metal braces, with fewer visits to the orthodontist required.

▓ A FIRST TOE IN THE WATER

Having recognized the opportunity and the threat that 3D printing presents, on paper at least, taking the next step can be daunting. With something so all-encompassing, so strategic, where to start? How to provide more impetus to what may have been a somewhat academic discussion? Key to answering that is to start small and build from there, usually with some sort of pilot project. This route offers a way to obtain tangible evidence of what 3D printing can do without significant risk to the normal day-to-day routines in a supply chain. It allows the people involved and affected to understand the possibilities and the constraints in a practical manner. Moreover, the current 3D printing sector facilitates this approach, allowing companies a "try before you buy" way of testing the technology using 3D printing hubs.

Of course, companies that have a more open approach to trialing new innovations can, and often do, simply acquire a 3D printer and let their supply chain and manufacturing teams play with it. That is a perfectly valid approach for many established companies with the available funding to do so. For other companies, particularly small and medium firms, or firms that are more risk averse, this approach may well be too expensive. A less risky and cheaper alternative is to outsource. 3D printer hubs and academic manufacturing departments offer 3D printing capability on a per-use basis without the expense of buying a printer and the ensuing expenditure on maintenance, training, and so on that comes with that. This is precisely the approach taken by Deutsche Bahn and embraced by the wider Mobility Goes Additive consortium of which it is part. The British utility company, Anglian Water, entered into a partnership with Sheffield University's Engineering Department to trial the manufacture of filter nozzles using 3D printing before using the same approach with other parts to explore the financial and engineering viability of the technology without buying a 3D printer themselves and committing to it. One of the greatest benefits of this approach is that its helps to drive momentum, using the early success stories to tangibly demonstrate what 3D printing can do. This helps the wider stakeholder community see what is possible and, with that knowledge, ignite imaginations and then identify other opportunities, recognizing the capabilities and restrictions of both 3D printing and the company's situation.

Of course, not every pilot will go on to grow. Supply chains may find that they have significant obstacles preventing them from adopting 3D printing, such as a lack of digital designs or inventory data, or that the technology is not right for their particular needs. Many issues can be resolved with their own improvement projects or separate efforts—designs can be digitized with 3D scanners or by hiring or outsourcing the necessary design skills. Inventories can be analyzed, with bills of materials (BOMs) data quality improved. With the advance of the digital supply chain, many of those will be needed sooner or later anyway.

That notwithstanding, in many cases, the companies that witness first-hand the benefits that 3D printing brings will then embark on a road toward adopting it, and that needs to be done in the right manner if it is to succeed.

 ## THE SUPERCHARG3D BLUEPRINT

The wider adoption of 3D printing in industrial situations is not simply a matter of installing a printer and making things. It is a strategic issue, something that involves many parts of a supply chain within an organization and, all too often, outside it as well. It requires leadership, communication, and teamwork, much as any other strategic change. There must not be any illusion that this is merely an operational or tactical decision. That starts by having the right group of people leading it: if 3D printing is to be more widely adopted in a supply chain, then a "coalition of the willing" is needed to continue to advocate for it, as mentioned earlier. This group, which should be led by a senior sponsor with sufficient vision and authority to drive the needed operational change, is typically made up of representatives of most of the affected areas of a business: manufacturing, design, supply chain, sales, and IT, with other areas augmenting the core team later. The aim here is to bring together those who can inform and shape the roadmap for adoption. Moreover, it ensures that all affected parts of a business have a say in how the exercise is delivered, something that is crucial to propelling it. In some companies, this group may well be headed by one or many of the most senior executives of the company, including the CEO. Generally, though, they will be capable leaders in each area, empowered to make decisions.

A structured and well-coordinated approach to adopting 3D printing is essential given the breadth of business areas that are impacted by its introduction and use. In broad terms, there are four key steps:

- Identify the key drivers that shape the business's objectives
- Evaluate the existing operations baseline
- Assess different solutions
- Operationalize the change across the company

These stages have been brought together to form the Supercharg3d Blueprint, a framework for the adoption of 3D printing in the supply chain, and it is depicted in Figure 9.2. A few attempts have been made to offer companies a guide on

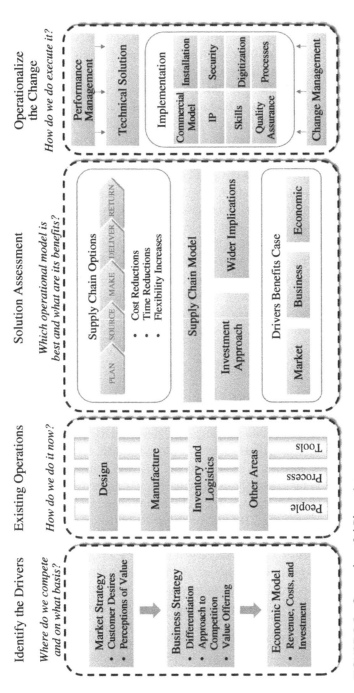

FIGURE 9.2 Supercharg3d Blueprint.

how to adopt 3D printing. However, all too frequently, these are too high level or too focused on production, not looking at the wider issues that need to be considered. Others fail to start with the most important question a company should ask: what do their customers want?

Instead, the Supercharg3d Blueprint aims to help supply chain directors and their companies approach the question of adopting 3D printing in a manner that addresses those issues head on. It is sufficiently detailed to make it relevant for most industry sectors without going too far into that detail to make it redundant; for instance, what is needed in a medical company will be different from the needs of an oil and gas company. At its heart is the implementation of 3D printing in the core areas of design and production but it also provides a wider systems-engineering view, considering the broader aspects that are essential if that implementation is to be sustainable. Much as Newton attributed his place in history by "standing on the shoulders of giants," so this framework has been informed by successful approaches for helping companies with strategic changes, and by lessons learned from the practical experience of dozens of companies of all sizes and across many sectors whose supply chains have adopted 3D printing. It looks at the key questions that need to be analyzed at each stage and in what order, providing guidance along the way.

IDENTIFY THE DRIVERS

No matter what they produce, companies exist to meet the requirements of their customers, be they external or internal to the firm. They have wants and needs, and they will turn to those firms that best meet them. They will decide on which option to choose based on what they see as the value that it brings, and that value can be one or several of many things. It may be simply price or minimum delivery time. It could be risk reduction, perhaps by purchasing spare parts to ensure that operations can be continued with minimal disruption. They may prefer quality and be willing to buy a high-spec part that has a lower risk of failure and a longer life. Alternatively, they may value a product that can be designed to their exact and unique specifications rather than having to adapt to it. Most likely, many of these factors will be combined to inform what a customer wants.

Normally, this information shapes the business strategy, addressing how a company differentiates itself from its competitors and stands out, and answering the questions of what it offers customers and how it does so. In other words, what is its value offer? Firms can compete based on many factors, providing value in price, service, responsiveness, choice, quality, or something else. Each part of that value offer shapes what the company needs to do to meet it, and the companies and supply chains that can best map their offer will succeed. Most companies know what drives their customers, even if they haven't written it down or articulated it directly. It can be found in the marketing strategy, in the sales approach used. The teams responsible for those are a good source to clarify what is needed. They will also be invaluable later when considering the value-adding benefits of 3D printing. The outputs from the sales and marketing teams feed the company's economic model, which brings together the expected revenues from customers, the costs of meeting those customers' needs, and the investment required to continue operations and grow. This model is then used to scale what the business wants and aspires to achieve, such as steady growth, the achievement of a particular revenue, a reduction in costs, or the maximization of the return on investment. Here too is a channel for identifying the value of meeting those customer needs. Is there more revenue to be gained by being faster? Do customers who get more customized products pay more for those? This analysis will identify the revenue drivers and help to focus on the tangible benefits of taking a different approach.

EVALUATE EXISTING OPERATIONS

Having identified the key drivers, the next step is to see how well a supply chain is doing to meet them, and where it could do better. As shown in the previous chapters, 3D printing can be used in many parts of the supply chain, each with their own benefits, costs, and broader consequences. This assessment of how that maps to what its customers want informs where 3D printing has the best role to play and where it can make the biggest impact. Ultimately, this will involve three elements of every productive supply chain:

1. Design, which articulates what is needed
2. Manufacture, which physically produces what is needed
3. Inventory and logistics, which gets what is needed to the right place at the right time

Design

Design is necessarily a function of purpose; what a part will be used for and how it will be used shape its form. The final design of an item is typically a compromise of customer requirements, often because the manufacture of lots of one becomes too costly and lengthy. Designers usually engage customers, and sometimes suppliers, for their feedback on designs, such as through prototypes, which can be expensive and take a long time to make. Moreover, although customer requirements may be dynamic and fast changing, the cycle time to understand and adopt those changes in designs tends to be prolonged. Last, the design can also be shaped by the restrictions of the manufacturing processes involved (e.g. perhaps several components need to be produced and assembled). Understanding the origin of the company's product design and what influences it helps to identify opportunities for 3D printing, both in form and process.

Good questions to examine include:

- How are my products designed?
- What design and manufacturing restrictions have to be contended with?
- How much variation is there across the products?
- How is the design process informed and what factors contribute to it?

Manufacture

The next step comes in understanding how products are made, both those manufactured by the company and the assemblies that it constructs from separate components. Traditional manufacture places several constraints on forms, driving limitations on what can be created and assembled. Complex designs are not traditionally viable, and allowances have to be made for the milling, cutting, filing, and other reductive steps. Those limitations are overcome by 3D printing. Mapping how each unit is made, down to the level of each of its component parts if it is assembled, and why that approach is used, helps identify opportunities for 3D printing. These may come from starting work more quickly or retooling production lines to cope with smaller design sets.

This part of the assessment needs to also look at the manufacturing process itself and the tools it uses. There may be opportunities to 3D print bespoke jigs and other elements of traditional manufacturing equipment. How are those sourced and made today? As mentioned in previous chapters, 3D printing allows for the faster production of injection molds, or the creation of molds with better heat dissipation and coolant channels.

It is also worth considering where manufacturing is located and why it is placed there. In many cases, space and cost are the critical drivers, and 3D printing may offer opportunities to reduce footprints. It could be that environmental or logistical requirements have been the drivers, keeping manufacturing close to raw materials, to power or water, to tax-incentivizing sites, and so on. The requirements and characteristics of 3D printing change that, enabling the conversion of advantages of placing manufacture closer to the client, such as opportunities in customer service and cost reductions, especially where international logistics are involved.

Good questions to examine will include:

- What manufacturing techniques are used to make things and why are those processes employed?
- What are the tool-up costs for new production, or retooling costs to change production?
- How long does it take to manufacture my parts and finished products?
- How quickly can I respond to changes in design, both large and small, and at what cost?
- What are my costs to manufacture a part?
- Can the manufacture be replaced or enhanced by 3D printing?
- Would 3D printing release the production of that item from the constraints of traditional approaches?
- How much can I improve service levels by moving manufacture closer to the end user?

Inventory and Logistics

Companies need to understand their own parts inventory; all too often, there is a reliance on out-of-date databases of the parts that they have in their inventory and of the components of those parts in often-superseded BOMs. The many forms of inventory that a company holds present various opportunities. Replacement parts for finished products, items to be used in manufacturing processes, maintenance, repair, and operations (MRO) spares, and works in progress (WIPs) are all typically held. The levels of those inventories are usually dictated by factors such as replenishment lead times, the service levels that items must have to meet their criticality, minimum and economic order quantities, and various other factors, each of which can be altered by 3D printing. Long tails of inventory, often needed for complex equipment with long replenishment or lifecycle times, are a good target for analysis. Each item in that long tail should be considered, looking at whether it is best stored as a digital file and printed on demand, and influencing that answer will be questions of design, materials, and location of manufacture. That analysis should consider the holding costs for the items throughout their expected lifecycle, and potential obsolescence costs if they aren't consumed. These costs then should be compared with the 3D-printed, "on demand" alternative, whether manufactured in-house or through a bureau, and recognizing the limitations that such an approach involves.

The characteristics of parts should also be examined. If there is variation in form between similar items or they need to be available in a variety of designs, then a 3D-printed solution may be more economic when viewed in terms of the whole cost and time to manufacture. More strategically, an examination of the current logistical footprint and what drives it should ask whether there are opportunities in moving manufacture closer to customers, like expediting international logistics and reducing transportation and import costs.

Good questions to examine will include:

- How do I distribute raw materials, WIPs, and finished products?
- What comprises my inventory?
- What proportion of it is in the long tail?
- Which items cause most stock outs?
- Which items have the longest lead times?
- What am I producing that isn't getting sold?
- What is my level of obsolescence?

In many cases, materials and spares will have been supplied to the company, so also understanding their source and involving suppliers in the analysis can reap benefits for both organizations. For instance, it is quite possible that the suppliers are themselves considering 3D printing options, and by working with their customers they can present a better offering. Such partnerships can also help customers resolve operational issues, such as reducing the risk of stock outs by employing a Manufacturing-as-a-Service approach.

The aim of this exercise is not to evaluate every possible item in the inventory—that would be far too onerous. Instead, it is to identify an initial set of items with which to start down the road to 3D printing, then implement a repeatable approach for evaluating new ones. Selecting the parts and items that will have the biggest impact on supply chain performance will prove the viability of the entire endeavor and demonstrate its value to the wider audience.

ASSESS DIFFERENT SOLUTIONS

This stage begins with identifying the options for using 3D printing in the company's supply chain, modeling those avenues while recognizing the need for investment and the wider implications, then building the business case for the best way forward. Will 3D printing be used to improve design or manufacture? Will using 3D printing in-house be better than outsourcing to a printing hub? Will 3D printing be carried out close to the customer? What does a shift from a make-to-stock to a make-to-order approach garner the company? Given the nature of the questions, this is best tackled top-down and bottom-up, starting from the strategic drivers that were previously identified to define the priorities in the supply chain. Those will inform where in the supply chain the focus should be. Typically, the considerations will be centered on changes to cost, time, or quality parameters, that will:

- Accelerate production
- Improve the flexibility of a manufacturing line
- Improve the design of products
- Respond to customer needs far more quickly

The assessment will be informed by the many benefits of 3D printing that have been presented in this book. Many approaches have been developed to translate and quantify those. For instance, an in-depth method for assessing the opportunity to reduce complexity in the supply chain was developed by Andre Kieviet in 2014 while he was at the University of Louisville.[6] Looking primarily at the reduction of complexity in the manufacturing element of the supply chain, Kieviet developed a clear approach that uses SCOR® supply chain parameters, such as Inventory Turnover, On Time Delivery, and Capacity Utilization. This is combined with the approach that Bernd Kaluza and his colleagues developed to measure supply chain complexity.[7] The result is an approach for analyzing the viability of 3D printing to reduce complexity and, through that, improve costs, production times, and so on such that it is beneficial to use it in that supply chain.

Needless to say, the assessment of different solutions will also consider the constraints that 3D printing still has, and the impacts of the wider implications that were described in Chapter 8. There is little point in embarking on a plan to purchase Selective Laser Sintering (SLS) 3D printers, for example, and use them in-house if there aren't the skills to use the equipment.

Each potential model will require a different investment strategy and, at this point, it is necessary to have a good view of the different potential technologies available, who can provide them, and what they can do, in order to map solutions to the model. That investment approach may involve establishing relationships with 3D printer manufacturers or other third-party hubs, or it may involve leasing or purchasing equipment and changing procurement to acquire the necessary raw materials. Adopting customer-managed inventory (CMI), for example, begins with establishing a close collaboration between the customer and its suppliers. First there needs to be an understanding of which items a company uses that can be 3D printed and which printers are best suited to make those. Next, the customer needs to work across its suppliers to identify commonality in 3D printers, picking out the best ones for its manufacturing needs. Customers then must agree how they will acquire or lease the 3D printers such that all parties benefit, and of course, customer staff will have to be trained in how to assign and use the printers, using Manufacturing Resource Planning (MRP) in more complex cases.

As well as the actual production of parts using 3D printing, the wider implications must be considered and any changes to the firm's operational model simulated to establish the consequences of the models being considered. For example, there may be a significant change in how the company deals with the digital designs, adapting legacy designs to the needs of 3D printing. It will include an assessment of the needs for expertise, from designers to 3D printer maintenance staff, identifying where those will come from and at what cost. There may be a need for additional finishing steps in the manufacturing process or for the introduction of technical or legal measures to protect designs and brands. Those considerations will naturally impact the timelines for using 3D printing, given the need to retrain or source staff and carry out the necessary changes.

Once the scenarios have been modeled, and the preferred solutions identified, it is then advisable to reverse the steps and consider the bottom-up implications: When using 3D printing, can other business drivers be improved? Can customers be offered a wider choice of designs at no extra cost? Is the company now able to deliver things that the existing models prevented it from doing? As Richard D'Aveni, the Bakala Professor of Strategy at Tuck University, says:

> "Leaders must consider the strategic implications as whole commercial ecosystems begin to form around the new realities of 3D printing"[8]

The results of these assessments then build up the business case, using the drivers of the business as a base and expressing the benefits that can be accrued based on investment in time, money, and resources to achieve them, as would be the case for any technology-based solutions.

This analysis of the company's ways of working and potential solutions will also serve to answer a critical question: When should 3D printing be trialed and adopted? While there isn't a formulaic answer to that question, the right answer is, "When the company is ready and able and when there is a tangible benefit in doing so." The output from the second and third phases just described will answer that question.

▦ OPERATIONALIZE THE CHANGE

While it would be great for everybody to be able to pick up a 3D printer and produce whatever they fancied, whenever they fancied, the reality is that things are not quite that simple. We have all become accustomed to buying an HP or Canon printer in Staples or Office World and printing off professional-looking documents in a matter of hours. 3D printing, however, is not so easy to use, particularly for industrial uses, just as any other newly installed significant manufacturing equipment would be.

With the business case agreed upon, implementing the chosen solution must be carefully controlled so as to not disrupt existing operations and impact the service that the firm provides to its customers. At the outset, this will mean ensuring that any changes in performance during the transition are identified early—much as civil engineers place instruments called Total Stations when building complex structures and tunnels to ensure that existing buildings aren't affected. This will include ensuring that the critical areas of the day-to-day business are monitored, typically through a dedicated and focused set of metrics and reporting tools. As far as is practicable, these should be leading indicators, which look at emerging trends and identify the issues before they materially affect the business, thereby allowing for any corrective actions to be taken during implementation.

The next priority is to define the technical solution and what it requires. That information will guide the procurement of the 3D printing solutions, whether leasing or acquiring 3D printers, purchasing materials, or contracting third-party service providers such as 3D printing hubs. If equipment needs to be installed, then there will be technical requirements, such as where machines will be sited, how they will be powered, and how they will integrate with existing materials flows.

The bulk of implementation, though, is a multifront endeavor, bringing together all the different functions that are affected, from manufacture and operations, through procurement and supply chain, to HR. As with any strategic operational decision, if it is to succeed, it needs to be something that the senior team in the firm endorse; therefore, it is very important that they buy into the business case and that the objectives of the exercise are agreed upon by all. Often in such situations, a "Tiger Team" approach is successful, bringing together managers who will guide the different parts of the business through implementation while aligning the various efforts with each other. In the case of 3D printing, that Tiger Team will have core tasks such as:

- *Research*: identifying the commercial models that will be changed and what the changes are (e.g. pricing, internally and externally).
- *Installation*: managing the physical introduction of any 3D printers.
- *IP*: ensuring that all protections are suitably covered, altering existing terms and agreements where necessary.
- *Security*: reviewing how design data is stored and used, who by, and where, to minimize the risk of sensitive data leaking.
- *Skills*: building the necessary capabilities in the company, from design through manufacture, to administer and run the 3D printing models; or outsourcing them. This will include, for instance, skills in Computer Aided Design design, the more technical skills required to set up and use the equipment, and skills required to finish a product after being 3D printed.
- *Digitization*: moving the flow of design and manufacturing data from paper-based to digital. In some cases, that can require the reverse digitization of designs, such as 3D-scanning parts, or converting paper drawings to digital drawings.
- *Quality Assurance*: reviewing what the necessary quality criteria are and developing the right processes, managerial and physical, to ensure that parts are made in accordance with the design requirements.
- *Processes*: underlining all the other areas, reviewing and realigning the processes throughout the supply chain to ensure that it works as intended and delivers the business case. For instance, processes for managing design changes, preparing prototypes, signing off on final designs, setting up and running 3D printers, and a myriad of other processes will need to be identified, checked and, where necessary, redrawn to take the new manufacturing processes and changes in supply chain into consideration.

Being clear on who does what throughout this process is very much an early imperative. The consideration and adoption of 3D printing is naturally a change program, and therefore the tools of change, such as process mapping and analytics, RACI matrices and terms of reference, must be developed early and updated often.

Identifying What to 3D Print

When Deutsche Bahn decided that 3D printing might be a preferable alternative to provide spare parts for its operations, they decided on a methodical, data-rich approach to identify which parts they should choose to make that way. The benefits of 3D printing an item can take many forms, from design to time to produce, and each of those will have some economic impact. Naturally, the expectation is that the analysis to identify suitable parts should be easy to complete, particularly in the modern operational environment with its vast pools of data. As Deutsche Bahn discovered, reality is not so simple.

First, there is a need to identify the set of items to be analyzed. Those items may be single objects or components for larger ones. They may be jigs and tools for use in manufacture, or the end-use items that are made. This requires understanding the list of materials that a company holds and uses, and the BOMs that make up assembled things. This data is typically found in materials management tools, product lifecycle management systems, and technical drawings. However, it is often incomplete, requiring an input from suppliers or carrying out a project to map the BOMs, neither of which is quick to do. Companies that have made public statements that they will be adopting 3D printing have found that the maturity and availability of this material information is woefully insufficient without considerable investment to clean up the datasets.

Once the items are identified, there is a need to understand economic factors, such as their total cost of ownership. This metric alone comprises several factors:

- Raw materials costs, whether base materials or parts to be assembled
- Production costs, including machinery, personnel, utilities, facilities, and so on
- Storage costs, including warehousing, insurance, and occasionally maintenance
- Logistics costs, including inspection, packaging and shipping, and transportation
- Depreciation costs, affecting items, and assets

The data required for these metrics goes beyond what is considered by the manufacturing team alone, and collaboration is therefore needed from the broader supply chain. Putting it together is not easy and requires inputs from ERPs, MRPs, and other operational and financial tools, together with an equitable allocation of costs.

As Deutsche Bahn and other companies have discovered, relying purely on data analysis to identify parts suitable for 3D printing requires

(continued)

(continued)

considerable time and investment and may not provide a conclusive answer. It isn't without merit, though. Approaches that have been used successfully have focused on providing indications of scale to hone in on the sets of items most suitable for 3D printing. For instance, items can be identified by physical criteria for suitability, such as whether they are made with single or multiple materials, whether they are assembled or singular. Together with the financial considerations, this will help to sift the range of items. However, it won't bottom out other opportunities, such as the simplification of composite items.

Table 9.1 offers a useful aide-mémoire to identify where 3D printing offers a good option. Brought together, the resultant analysis will identify the best supply chain model to employ, from using 3D printing to parallel production to a complete redesign of the customer-supplier arrangements.

The best approach is to convene a group from the manufacturing and supply chain teams. Informed by the data analysis, this team can use their knowledge and experience to identify suitable items to 3D print, bridging the intangible gaps in the assessment. They can ensure that cost comparisons with traditional manufacturing are made based on acceptable assumptions and approaches, considering the right variables such as manufacturing lead times and inventory holding costs. Moreover, they can ensure the balanced consideration of quality, cost, and time benefits.

The extent of the plan for integrating 3D printing will naturally be a function of the scale of the required change, and implementing it needs to be administered as all significant change management projects are. This includes establishing the goals for the project, its short- and long-term targets, embedding the necessary governance and, above all, communicating with all those in the company and beyond who are affected at the right time and in the right manner. This latter element, that of organizational change, is very often the hardest to tackle. All too often, changes to working routines and workforce sizes must be negotiated with government and union bodies, and if this is done badly, then such change initiatives—which the adoption of 3D printing can quickly become—will stall and very possibly fail. Therefore, planning for that organizational change should be viewed as a critical success factor for the use of 3D printing.

TABLE 9.1 Guidelines on 3D printing applicability.

Characteristic	High	Medium	Low
Required volumes	No	Maybe	Yes
Speed of response	Yes	Maybe	No
Product cost	Maybe	Yes	Maybe
Level of customization	Yes	Yes	Maybe
Range size	Yes	Yes	Maybe
Number of materials	No	Maybe	Yes

Change Management

As was stated at the outset of this chapter, adopting 3D printing is a strategic activity, and the Supercharg3d Blueprint offers a practical roadmap for that. It includes the granular analysis of the organization, from its strategy to its supply chain. If using it reveals that 3D printing offers a net benefit to the company's performance, then there will be a period of planning and implementation of significant changes to the structures, processes, skills base, and tools. In the fourth stage of the roadmap is the element Change Management, which is where those steps are implemented.

There is a whole industry of books and thought leadership on the topic of change management, and the reader is encouraged to seek out those whose philosophies best align with the corporate culture. There are some common threads to the leading approaches that are especially relevant to the Supercharg3d Blueprint and that need to be part of the entire approach from its inception: communication and listening, the need for quick wins, advancing in steps, and flexibility with an eye on the end goal. Core to this is a need to have visibility of what will change, why it will change, and how it will change, even when these may be developing, with the aim of gaining and maintaining cooperation from all involved throughout the endeavor. Without strong change management, any effort to adopt 3D printing will be destined to fail.

Approaching the evaluation of whether 3D printing is right for a business and, if so, of adopting it significantly transforms that business's potential capability. The realization of that potential doesn't stop once the capability is deployed; rather it evolves as the business and its supply chain get used to what it can do and begin to change their supply chain model. By doing that, the significant competitive differentiators of fast production cycles, quicker delivery, increased flexibility in design, and so on can be achieved. Several major manufacturers have recognized its potential and, over the last few years, have entered into strategic partnerships with the makers of 3D printers or acquired the leading firms. In 2016, the engineering firm GE, which had already established a name for itself as a leading developer as well as a user of 3D printers, announced that they were acquiring three companies, Sweden's Arcam AB, and Concept Laser and SLM Solutions Group AG from Germany. Having recognized the advantages of 3D printing, GE is now positioning itself further as a premier manufacturer, using the technology to give them that competitive advantage: making better products and making products better. This move is also changing their supply chain more holistically, opening new models and changing the dynamics and operations of the whole company. These achievements, however, are not the preserve of global firms like GE: one of the greatest advantages of 3D printing to all sizes of company and supply chain is that it is eminently scalable. All that is needed is a willingness to make the change and to approach it in the right manner. This book helps to achieve those.

 ## Notes

1. Wohlers Associates, *Wohlers Report 2018*, https://wohlersassociates.com/2018report.htm.
2. John P. Kotter, *Leading Change* (Boston: Harvard Business Review Press, 1996).
3. Basavaraj S. Phulari, *History of Orthodontics* (New Delhi: Jaypee Brothers Medical Publishers, 2013).
4. André Kieviet and Suraj M. Alexander, "Is Your Supply Chain Ready for Additive Manufacturing?" *Supply Chain Management Review*, October 16, 2017, http://www.scmr.com/article/is_your_supply_chain_ready_for_additive_manufacturing.
5. T. J. McCue, "3D Printing Moves Align Technology Toward $1.3 Billion In Sales," *Forbes*, September 14, 2017, https://www.forbes.com/sites/tjmccue/2017/09/14/3d-printing-moves-align-technology-toward-1-3-billion-in-sales/.
6. André Kieviet, "Implications of Additive Manufacturing on Complexity Management within Supply Chains in a Production Environment" (PhD diss., University of Louisville, 2014).
7. B. Kaluza, H. Bliem, and H. Winkler, "Strategies and Metrics for Complexity Management in Supply Chains," in *Complexity Management in Supply Chains*, T. Blecker and W. Kersten, eds. (Berlin: Erich Schmidt, 2006).
8. Richard D'aveni, "The 3-D Printing Revolution," *Harvard Business Review*, May 2015, 40–48.

Epilogue

ILL GATES IS credited with saying that we overestimate what is possible in one year and underestimate what can be accomplished in 10. Yes, 3D printing has come out of a multiyear hype cycle, but that hasn't held it back. With the advent of digital transformation, the increased use of automation, and the focus on addressing the issues that have long hampered its widespread adoption, 3D printing is now on a course to realize the promise that its disruptive nature has long held.

In late 2015, DHL published its vision of how supply chains will look by 2025. They concluded that end users will continue to demand more customization, meaning that supply chain managers will have to cope with continually increasing complexity, and therefore they must deliver an ever-broader portfolio of products. In parallel, markets will continue to change their dynamics, with more suppliers emerging, a growing lacuna of qualified and skilled people, and new technologies emerging to bridge that situation. The result, DHL made clear, is that companies will be forced to rethink their current supply chain models. Future supply chains will be digital, networked, transparent, collaborative, intelligent, and resilient. Those qualities are all realized with 3D printing. As Barrett Thompson at Zilliant said:

> "Viable use cases for 3D printing technology are rapidly growing, and as they do, industrial B2B manufacturers and distributors will be pressed to find creative ways to incorporate them."[1]

Discussion of 3D printing has moved from laboratories and prototype makers to manufacturing shop floors and boardrooms. In 2015, over 30% of the top 300 largest global brands were using 3D printing or actively evaluating how to do so,[2] growing to 81% in 2017 according to the manufacturing services firm Jabil.[3] Ninety percent of companies that have adopted 3D printing see it as delivering a competitive advantage.[4] Avi Reichental, the former President and CEO of 3D Systems, is quite bullish, saying in 2018:

> "In a decade it will be ubiquitous in industry, architecture, education, and the engine behind a new generation of digital artisanry."[5]

Truly, 3D printing is now and growing in importance, and companies must react accordingly. If history has taught us one thing, from the invention of the assembly line to the introduction of data management, it is that the first movers are most likely to succeed in any new paradigm. Of course, investing in a new technology and adapting the supply chain and operations of a company is a big gamble. However, the rise of 3D printing service providers and lease models, and the increase in digital technologies in manufacturing, all help to facilitate the use of 3D printing and mitigate the risk.

Advances in 3D printing technology cover all its value chain. The range of materials is expanding faster than at any time since the technology was invented in the 1980s; the variety of techniques and the speeds, precision, and tolerances that they deliver are continually improving, driven by the explosion of start-ups and research efforts over recent years (which makes writing about them in a book a challenge of relevance!). For instance, within 24 months after it first demonstrated its Continuous Liquid Interface Production technology, Carbon3D announced a commercial arrangement with Siemens and Adidas to make sneaker midsoles inside 90 minutes using it. Those advances are already being banked on for truly ambitious developments: the NASA-led operational model for the establishment of permanent settlements on the Moon and thereafter Mars involves using the technology to make the buildings needed using lunar and Martian materials, and potentially to provide the food for the colonists using feedstock sent up in advance or with them. The technology is already being tested, and companies in Dubai, for instance, are already producing entire structures using concrete and related materials, from offices to residences. Even where one's company is not necessarily leveraging 3D printing directly, the effects of competitors, suppliers, and customers using it present a risk. As aircraft manufacturers use the benefits of topology optimization and 3D printing to lighten the loads on their airplanes, so the demand for fuel per journey drops, which affects the revenues of the oil and gas companies that supply that fuel.

Tomorrow's managers are today's students, and 3D printing is now entering the minds of those who will decide on how to make things tomorrow. Over 200 universities and colleges already offer 3D courses in their curricula, covering aspects not only of 3D printing but also 3D scanning and design. As T. J. McCue told the *Harvard Business Review*:

> "For me, the most important 'tipping point' isn't about how many manufacturers have changed, it's about how many minds have. Thanks to more accessible technology, we are now reaching a critical mass of people who, when they think about how things are made, think in a different way. You could say they are thinking in 3D."[6]

The flexibility it brings opens possibilities for small businesses like never before, promising increased innovation in design and new products. Those will be unrestricted by complexity, have fewer parts, and employ more feedback sooner than has been achievable so far. Key to that is understanding that there is a new paradigm and that anything is possible. Henry Ford understood this: it was by recognizing that there was a new way of making things, and by using that new capability that he was able to change the way that things were made—and how the world works—so radically. Much as designs were transformed, supply chain models were changed, with significantly different dynamics and relationships between suppliers and customers. Likewise, 3D printing changes supply chains—how will you adapt yours?

Notes

1. Barrett Thompson, "How 3D Printing Stands to Disrupt Capacity, Sourcing and Pricing," *Supply & Demand Chain Executive*, January 4, 2016, http://www.sdcexec.com/article/12154349/how-3d-printing-stands-to-disrupt-capacity-sourcing-and-pricing.
2. T. J. McCue, "3D Printing Is Changing the Way We Think," *Harvard Business Review*, July 21, 2015, https://hbr.org/2015/07/3d-printing-is-changing-the-way-we-think.
3. Jabil,. "3D Printing Trends Show Positive Outlook," November 6, 2017, https://www.jabil.com/insights/blog-main/3d-printing-trends-show-positive-outlook.html.
4. Sculpteo, "The State of 3D Printing 2017," accessed September 25, 2018, https://www.sculpteo.com/en/get/report/state_of_3D_printing_2017/.
5. expertinsights, "The 3D Printing Influencer Series Part 4: Avi Reichental," Disruptordaily.com, May 23, 2018, https://www.disruptordaily.com/the-3d-printing-influencer-series-part-4-avi-reichental.
6. McCue, "3D Printing."

Appendix: Definition of Supply Chain Metrics

THE APICS SUPPLY Chain Council has developed and maintained the Supply Chain Operations Reference (SCOR) model to measure cross-functional, cross-company supply chain processes. Its Level 1 metrics measure five supply chain performance attributes:

- Supply Chain Reliability, referring to the ability of the supply chain to perform tasks as expected. Reliability focuses on the predictability of the outcome of a process; its metrics concern time, quantity, and quality.
- Supply Chain Responsiveness refers to the speed at which tasks are performed within the supply chain, particularly the speed at which a supply chain provides products to the customer. Responsiveness metrics include those concerned with cycle times.
- Supply Chain Agility, which is the ability to respond to external influences, such as changes that gain or maintain a competitive advantage. This includes the ability to respond to marketplace changes to gain or maintain a competitive advantage. Agility metrics typically measure the flexibility and adaptability of the supply chain.
- Supply Chain Costs, referring to the cost of operating the span of supply chain processes. This will include costs associated with labor and resources, materials, transportation, and the management of the parts and entirety of the supply chain.
- Supply Chain Asset Management, which is a more recent addition to the SCOR attributes and looks at the ability of the supply chain to use assets efficiently. Strategies include inventory optimization and insourcing versus outsourcing, with metrics such as Inventory Days of Supply and Capacity Utilization.

The metrics in this book include many of those within the SCOR framework, and a few go beyond those. For consistency and clarity, Table A.1 presents the definitions of the individual metrics used in the book.

TABLE A.1 Definition of supply chain metrics.

Metrics Name	Definition
Carbon Footprint	A measure of the carbon dioxide emissions resultant from establishment and operation of the supply chain.
Cost of Obsolescence	The cost of materials and inventory that are disposed of as obsolete.
Cost of Warranties	The cost of servicing warranties and guarantees on products and parts made and supplied by the supply chain its end users.
Delivery Lead Times	The time from the receipt of a customer order to the delivery of the product.
Delivery Performance to Customer Commit Date	The ratio of orders successfully delivered by the customer commit date.
Disposition Costs	The cost of disposals, including obsolescence, landfill, etc.
Fulfillment Costs	The costs associated with fulfilling an order, including transportation, customs, duties, taxes, and tariffs, labor, automation, property, plant and equipment, and inventory and overhead costs.
Inventory Costs	The cost of inventory held.
Inventory Days of Sale	The average value of inventory divided by cost of goods sold
Make Cycle Time	The time to make a product, including production engineering, issue of material, production and testing, the release of finished products, production activity scheduling, and product finishing and packaging.
Order Fulfillment Cycle Time	The time to fulfill an order, including sourcing, making and delivering it.
Perfect Order Fulfillment	The measure of how many orders are completely and correctly fulfilled, including orders delivered in full, meeting customer commit dates, providing accurate documentation (if relevant), and delivering the products in perfect condition.
Percentage of Orders Delivered in Full	The ratio of orders that are delivered correctly in terms of what they are and how many they are.
Production Cost	The costs associated with producing a product, including labor, automation, property, plant and equipment, and inventory and overheads.
Retooling Time	The costs incurred in retooling a production line to cope with changing orders.
Sourcing Cost	The costs associated with sourcing materials and services to make and deliver a product, including labor, automation, property, plant and equipment, and inventory and overheads.

TABLE A.1 Definition of supply chain metrics. (*continued*)

Metrics Name	Definition
Supply Chain Flexibility	The number of days required to achieve an unplanned sustainable 20% increase/decrease in quantities delivered.
Supply Chain Adaptability	The measure of the quantity of increased/decreased production a supply chain can achieve and sustain in 30 days' time.
Supply Chain Risk	A measure of the risk in the supply chain, typically a rating based on an overall score to summarize it qualitatively (from very low to very high) and/or quantitatively (from 1 to 25, say).
Total Cost to Serve	The total cost to serve a customer's order, including planning, sourcing, material landing, production, order management, fulfillment and returns costs, and the cost of goods sold.
Transportation Costs	The costs involved in moving materials, parts, and products along and within the supply chain.
Wastage Rate	The percentage of materials wasted during manufacture.

Index

Note: Page references in *italics* refer to figures and tables.